Technical Mathematics with Calculus

Technical Mathematics with Calculus

Editor

Abby Enger

scitus academics

Technical Mathematics with Calculus
Edited by **Abby Enger**

ISBN: 978-1-68117-650-5
Library of Congress Control Number: 2016946516

SCITUS Academics LLC,
616, Corporate Way, Suite 2, 4766,
Valley Cottage, NY 10989

www.scitusacademics.com

Notice

Printed in United States of America on Acid Free Paper

Preface

Mathematics is the science that deals with the logic of shape, quantity and arrangement. Math is all around us, in everything we do. It is the building block for everything in our daily lives, including mobile devices, architecture (ancient and modern), art, money, engineering, and even sports. Mathematics is a universal science language that makes use of symbols and notations for describing numerical, geometric and graphical relationships. It is a human activity that involves observing, representing and investigating patterns and qualitative relationships in physical and social phenomena and between mathematical objects themselves. It helps to develop mental processes that enhance logical and critical thinking, accuracy and problem solving that will contribute in decision-making. Mathematical problem solving enables us to understand the world around us, and, most of all, teaches us to think creatively. The aim of Technical Mathematics is to apply the Science of Mathematics to the Technical field where the emphasis is on application and not on abstract ideas.

Technical Mathematics with Calculus focuses on developing students' critical thinking skills as well as improving their proficiency in a broad range of technical math topics such as algebra, linear equations, functions, and integrals. It will be of valuable to advanced graduate students, lecturers as well as practitioners.

Table of Contents

1

CHAPTER

Base-X Notation and Tri-Value Logic

Xiaolin Li

Department of Mathematics and Computer Science, Alabama State University, Montgomery, USA

ABSTRACT

This paper introduces the base-X notation and discusses the conversion between numbers of different bases. It also introduces the tri-value logic that is associated with the base-3 system.

Keywords: Index Terms—Base-X Notation; Tri-Value Logic

1. INTRODUCTION

People use decimal numbers in their daily life. However, our computers use the binary system and the binary logic [1-3]. A computer also involves other number systems such as octal numbers and hexadecimal numbers. There is an easy way to convert a number in one system to another using the base-X notation.

2. BASE-X NOTATION

Let X be an integer, X > 1. Using X as the base, we can form a number system. A base-X system uses the digit set {0, 1, ···, X − 1} to constitute a number. Any number Y in the system can be written as:

$$Y = Y_m Y_{m-1} \cdots Y_1 Y_0 \cdot Y_{-1} Y_{-2} \cdots Y_{-n} \tag{1}$$

where m and n are integers, 0 ≤ m <∞, 1 ≤ n < ∞. In the above notation, Y_i is the digit at position i and Y_i is in {0, 1, ···, X − 1}, −n ≤ i ≤ m. When X = 2, we have the binary system; when X = 8, we have the octal system; when X = 10, we have the decimal system; when X = 16, we have the hexadecimal system.

Let us consider how to convert Y into a decimal number when X ≠ 10. Notice that for each position i, there is an associated weight X^i, and Y_i represents a term Y_iX^i. Thus, adding all terms together, we get the corresponding decimal number:

$$\begin{aligned} Y &= Y_m Y_{m-1} \cdots Y_1 Y_0 \cdot Y_{-1} Y_{-2} \cdots Y_{-n} \\ &= Y_m X^m + Y_{m-1} X^{m-1} + \cdots + Y_1 X + Y_0 + Y_{-1} X^{-1} \\ &\quad + Y_{-2} X^{-2} + \cdots + Y_{-n} X^{-n} \\ &= \sum_{i=-n}^{m} Y_i X^i \end{aligned} \tag{2}$$

The following examples illustrate the above conversion, where the subscript indicates the base:

1) Binary to Decimal

$$(11001.101)_2 = \left(1\times2^4 + 1\times2^3 + 1 + 1\times2^{-1} + 1\times2^{-3}\right)_{10}$$
$$= \left(25\frac{5}{8}\right)_{10}$$

The binary system has the advantage that it uses only two digits: 0 and 1. Within a binary number, digit 0 always represents a zero term while digit 1 always represents a term identical to the weight associated with the digit position. Notice that we have dropped all zero terms in the above summation.

2) Octal to Decimal

$$(756.23)_8 = \left(7\times8^2 + 5\times8 + 6 + 2\times8^{-1} + 3\times8^{-2}\right)_{10}$$
$$= \left(494\frac{19}{64}\right)_{10}$$

3) Hexadecimal to Decimal

$$(5FE\cdot2B)_{16}$$
$$= \left(5\times16^2 + 15\times16 + 14 + 2\times16^{-1} + 11\times16^{-2}\right)_{10}$$
$$= \left(1534\frac{43}{256}\right)_{10}$$

The digit set used in the hexadecimal system is {0, 1, 2, 3, 4, 5, 6, 7, 8, 9, A, B, C, D, E, F}, where A = 10, B = 11, C = 12, D = 13, E = 14, F = 15.

3. DECIMAL TO BASE-X CONVERSION

The conversion of a decimal number to a base-X number is also indicated by Equation (2). Given a decimal number Y, we can treat Y as the summation of a sequence of terms where each term is a power of X, then we can write down the corresponding base-X number. For example, given the decimal number 25.625, we have

$$(25.625)_{10} = \left(16+8+1+\frac{1}{2}+\frac{1}{8}\right)_{10}$$

therefore, $(25.625)_{10} = (11001.101)_2$. In general, one can use the following procedure for the conversion.

Let $Y = Y_i \cdot Y_f$, where Y_i and Y_f are the integer portion and fraction portion of Y, respectively. We can convert Y_i and Y_f separately:

1) Regarding Y_i as the initial quotient and dividing the remained quotient by X repeatedly until it becomes zero, we get a sequence of remainder digits, the integer portion of the base-X number then can be obtained by reversing the order of the remainder digits in the sequence;

2) Regarding Y_f as the initial fraction and multiplying the remained fraction by X repeatedly, we get a sequence of digits from the integer portion of the products, the fraction portion of the base-X number then can be obtained from this digit sequence.

Consider the decimal number 123.65625 for example, we can convert it into binary, octal and hexadecimal number:

1) Decimal to Binary Integer portion conversion:

Quotient	Remainder
123	
61	1
30	1
15	0
7	1
3	1
1	1
0	1

Fraction portion conversion:

Fraction	Integer Product
0.65625	
0.3125	1
0.625	0
0.25	1
0.5	0
0	1

$$(123.65625)_{10} = (1111011.10101)_2$$

2) Decimal to Octal Integer portion conversion:

Quotient	Remainder
123	
15	3
1	7
0	1

Fraction portion conversion:

Fraction	Integer Product
0.65625	
0.25	5
0	2

$$(123.65625)_{10} = (173.52)_8$$

3) Decimal to Hexadecimal Integer portion conversion:

Quotient	Remainder
123	
7	*B*
0	7

Fraction portion conversion:

Fraction	Integer Product
0.65625	
0.5	*A*
0	8

$$(123.65625)_{10} = (7B \cdot A8)_{16}$$

Notice that in the above conversion, we always get a finite digit sequence for the integer portion because the remained quotient always becomes zero by repeated division. On the other hand, we may get an infinite digit sequence for the fraction portion if the remained fraction never becomes zero through the repeated multiplication. In such a case, it may produce an endless circulate fraction.

For example, $(0.4)_{10} = (0.011001100110\cdots)_2$.

4. BASE-3 SYSTEM AND TRI-VALUE LOGIC

Let us consider the base-3 number system that uses the digit set {0, 1, 2} to constitute a number.

Denote T = {0, 1, 2} and let A, B in T, the results of A + B are shown in Table 1.

Consider a tri-value logic associated with the digit set T. Let 0, 1, 2 represent false, true, and neutral (neither false nor true) state, respectively. We can generalize the basic logic operators $\cap$ (logic and), $\cup$ (logic or) and ′(logic not) in Boolean algebra to this tri-value case such that $\cap : T^2 \rightarrow T$, $\cup : T^2 \rightarrow T$, and $' : T \rightarrow T$. Let A and B be two variable, A, B in T, then the definition of these generalized logic operators is given in Table 2.

Notice that if we delete all rows that contain state 2 from Table 2, then we get the truth table in Boolean algebra as Table 3.

From Table 2, it is easy to see that the following equalities hold:

$$A \cap A = A$$

$$A \cup A = A$$

$$0 \cap A = 0$$

$$0 \cup A = A$$

$$1 \cap A = A$$

$$1 \cup A = 1$$

It is obvious that the De Morgan's law and the binary implication can be extended into this tri-value case:

$$(A \cap B)' = A' \cup B'$$

$$(A \cup B)' = A' \cap B'$$

$$A \rightarrow B = A' \cup B$$

In addition to the above, we can also define other compound logic operators from the basic logic operators $\cap$, $\cup$, ', such as the exclusive or operator $\oplus : T^2 \rightarrow T$:

$$A \oplus B = A \cap B' \cup A' \cap B$$

Table 1. Base-3 addition.

A/B	0	1	2
0	0	1	2
1	1	2	10
2	2	10	11

Table 2. Tri-value logic.

A	B	$A \cap B$	$A \cup B$	A'	B'	$A' \cap B'$	$A' \cup B'$
0	0	0	0	1	1	1	1
0	1	0	1	1	0	0	1
0	2	0	2	1	2	2	1
1	0	0	1	0	1	0	1
1	1	1	1	0	0	0	0
1	2	2	1	0	2	0	2
2	0	0	2	2	1	2	1
2	1	2	1	2	0	0	2
2	2	2	2	2	2	2	2

Table 3. Binary logic.

A	B	$A \cap B$	$A \cup B$	A'	B'	$A' \cap B'$	$A' \cup B'$
0	0	0	0	1	1	1	1
0	1	0	1	1	0	0	1
1	0	0	1	0	1	0	1
1	1	1	1	0	0	0	0

5. CONCLUSION

In this paper, we have introduced the base-X notation and have discussed the conversion between numbers of different bases. We have also introduced a tri-value logic that is associated with the base-3 system. We have shown that the tri-value logic is compatible with the binary logic, and the De Morgan's law and the implication rule can be extended into this tri-value case. Here we point out that all base-X systems are equivalent to each other because any number in one system can be uniquely mapped into another system as implied by Equation (2). Furthermore, any rational number in one system remains rational in another system because its integer numerator and denominator can be converted into integers in another system as indicated by the integer conversion process.

REFERENCES

1. J. Sanchez and M. P. Canton, "Microcontroller Programming: The Microchip PIC," CRC Press, Boca Raton, 2006.
2. J. E. Whitesitt, "Boolean Algebra and Its Applications," Dover Publications, New York, 2010.
3. S. Givant and P. Halmos, "Introduction to Boolean Algebras (Undergraduate Texts in Mathematics)," Springer, New York, 2008.

2

CHAPTER

A Simple Way to Prove the Characterization of Differentiable Quasiconvex Functions

Giorgio Giorgi

Faculty of Economics, University of Pavia, Pavia, Italy

ABSTRACT

We give a short and easy proof of the characterization of differentiable quasiconvex functions.

Keywords: Quasiconvex Functions, Generalized Convexity

1. INTRODUCTION

Quasiconvex functions play an important role in several branches of applied mathematics (e.g. mathematical programming, minimax theory, games theory, etc.) and of economic analysis (production theory, utility theory, etc.). De Finetti [1] was one of the first mathematicians to define quasiconvex functions as those functions $f : X \to \mathbb{R}$, $X \subset \mathbb{R}^n$ being a convex set, having convex lower level sets, i.e. the set

$$L(f,\alpha) = \{x : x \in X, f(x) \le \alpha\}$$

is convex for every $\alpha \in \mathbb{R}$.

De Finetti did not name this class of functions: the term "quasiconvex (quasiconcave) function" was given subsequently by Fenchel [2] . It is well-known that the above characterization is equivalent to

$$x^1, x^2 \in X,\ f(x^2) \le f(x^1) \Rightarrow f(\lambda x^1 + (1-\lambda)x^2) \le f(x^1),\ \forall \lambda \in [0,1],$$

i.e., in a more symmetric way,

$$f(\lambda x^1 + (1-\lambda)x^2) \le \max\{f(x^1), f(x^2)\},\ \forall \lambda \in [0,1],\ \forall x^1, x^2 \in X.$$

When f is differentiable on the open convex set $X \subset \mathbb{R}^n$, we have the following characterization of a quasiconvex function.

Theorem 1. Let $f : X \to \mathbb{R}$ be differentiable on the open convex set $X \subset \mathbb{R}^n$. Then f is quasiconvex on X if and only if

$$x^1, x^2 \in X, f(x^2) \le f(x^1) \Rightarrow \nabla f(x^1)(x^2 - x^1) \le 0. \tag{1}$$

Theorem 1 was given by Arrow and Enthoven [3] ; however, these authors prove, in a short and easy way, only the necessary part of the theorem, but not the converse property, whose proof is indeed presented in a quite intricate way by several authors (see, e.g. [4] -[11]). Here we present an easy proof of Theorem 1, by exploiting some results on quasiconvexity of functions of one variable, results therefore suitable for geometrical illustrations. We need two lemmas.

Lemma 1. Let $f: X \to \mathbb{R}$, $X \subset \mathbb{R}^n$ a convex set. Then f is quasiconvex on X if and only if the restriction of f on each line segment contained in X is a quasiconvex function, i.e. if and only if the function $\theta(t) = f(x+ty)$ is quasiconvex on the interval

$$I_{x,y} = \{t \in \mathbb{R} : x+ty \in X\}.$$

Proof. The quasiconvexity of $\theta(t)$ is equivalent to the implication

$$t_1, t_2 \in I_{x,y},\ \theta(t_2) \le \theta(t_1) \Rightarrow \theta(t_1 + \lambda(t_2 - t_1)) \le \theta(t_1),\ \forall \lambda \in [0,1].$$

By setting $x^1 = x + t_1 y,\ x^2 = x + t_2 y,$ we have $x^1 + \lambda(x^2 - x^1) = x + t_1 y + \lambda(t_2 - t_1) y.$ The thesis follows by noting that $\theta(t_2) \le \theta(t_1)$ and the logical implication $\theta(t_2) \le \theta(t_1) \Rightarrow \theta(t_1 + \lambda(t_2 - t_1)) \le \theta(t_1)$, $\forall \lambda \in [0,1]$ are equivalent to $f(x^2) \le f(x^1)$, and $f(x^1 + \lambda(x^2 - x^1)) \le f(x^1)$, $\forall \lambda \in [0,1]$, respectively.

The next lemma is proved in Cambini and Martein [12] and is given also by Crouzeix [13] , without proof.

Lemma 2. Let $\varphi : I \to \mathbb{R}$ be differentiable on the interval $I \subset \mathbb{R}$; then φ is quasiconvex on I if and only if

$$t_1, t_2 \in I,\ \varphi(t_2) \leq \varphi(t_1) \Rightarrow \varphi'(t_1)(t_2 - t_1) \leq 0. \quad (2)$$

Proof. Let $t_1, t_2 \in I$ such that $t_1 < t_2 \,(t_1 > t_2)$ and $\varphi(t_1) \geq \varphi(t_2)$. The quasiconvexity of φ implies $\varphi(t) \leq \varphi(t_1),\ \forall t \in [t_1, t_2]$ $(\forall t \in [t_2, t_1])$ so that φ is locally non-increasing (locally non-decreasing) at t_1 and consequently (2) holds. Assume now that (2) holds. If φ is not quasiconvex, there exist $t_1, t_2 \in I$ with $t_1 < t_2$

such that $M = \max\{\varphi(t), t \in [t_1, t_2]\} > \max\{\varphi(t_1), \varphi(t_2)\}$.

Let $\bar{t} = \inf\{t \in [t_1, t_2] : \varphi(t) = M\}$; the continuity of

φ implies the existence of $\varepsilon > 0$ such that $\varphi(t) < M$, $\forall t \in (\bar{t} - \varepsilon, \bar{t})$ and $\varphi(t) > \max\{\varphi(t_1), \varphi(t_2)\}$,

$\forall t \in (\bar{t} - \varepsilon, \bar{t})$. The mean value theorem applied to the interval $[\bar{t} - \varepsilon, \bar{t}]$ implies the existence of

$t^* \in (\overline{t} - \varepsilon, \overline{t})$ such that $\varphi'(t^*) > 0.$ Consequently, we have $\varphi(t^*) > \varphi(t_2)$ with $\varphi'(t^*) > 0$ and this contradicts (2).

Proof of Theorem 1.

It is sufficient to note that if f is differentiable on the open and convex set $X \subset \mathbb{R}^n$, then we have, with $\theta(t) = f(x + ty)$, $\theta'(t) = \nabla f(x + ty)y.$

Therefore, on the ground of Lemmas 1 and 2, if we put $x^1 = x + t_1 y,\ x^2 = x + t_2 y,$ we have $f(x+t_2 y) \le f(x+t_1 y) \Rightarrow \nabla f(x+t_1 y)y(t_2 - t_1) \le 0,$ i.e. ■
$f(x^2) \le f(x^1) \Rightarrow \nabla f(x^1)(x^2 - x^1) \le 0.$ $x^1, x^2 \in X,$

Finally, we point out that the proof of Ponstein [10] can be shortened as follows:

1) Let f be quasiconvex (and differentiable) on the open and convex set $X \subset \mathbb{R}^n$, i.e. let

$$x^1, x^2 \in X,\ f(x^2) \le f(x^1) \Rightarrow f(\lambda x^1 + (1-\lambda)x^2) \le f(x^1),\ \forall \lambda \in [0,1].$$

By the mean value theorem there exists a number $\xi \in (0,1)$ such that

$$\nabla f\left(x^1+\xi\lambda\left(x^2-x^1\right)\right)\lambda\left(x^2-x^1\right)=f\left(x^1+\lambda\left(x^2-x^1\right)\right)-f\left(x^1\right)\le 0.$$

Dividing by $\lambda(>0)$ and letting $\lambda \to 0,$ we have $\nabla f\left(x^1\right)\left(x^2-x^1\right)\le 0.$

2) Assume conversely that (1) holds and that for $x^1, x^2 \in X,$ with $f\left(x^2\right)\le f\left(x^1\right)$ there exists a point x^3 between x^1 and x^2 with $f\left(x^3\right)>f\left(x^1\right).$ Then there exists near x^3 (e.g. between x^1 and x^3) also a point x^4 with $f\left(x^4\right)>f\left(x^1\right)$ and $\nabla f\left(x^4\right)\left(x^2-x^4\right)>0.$ Indeed, if for all x between x^1 and x^3 it would hold $\nabla f(x)\left(x^2-x\right)\le 0$ or $f(x)\le f\left(x^1\right),$ then we would have $f\left(x^3\right)\le f\left(x^1\right)$ but not $f\left(x^3\right)>f\left(x^1\right).$ But, being $f\left(x^2\right)<f\left(x^4\right)$, this implies $\nabla f\left(x^4\right)\left(x^2-x^4\right)\le 0$, in contradiction with the inequality previously obtained.

REFERENCES

1. De Finetti, B. (1949) Sulle Stratificazioni Convesse. Annali di Matematica Pura ed Applicata, 30, 173-183.

2. Fenchel, W. (1953) Convex Cones, Sets and Functions. Lecture Notes, Princeton University, Princeton.

3. Arrow, K.J. and Enthoven, A.C. (1961) Quasi-Concave Programming. Econometrica, 29, 779-800.

4. Avriel, M. (1976) Nonlinear Programming: Analysis and Methods. Prentice-Hall, Englewood Cliffs.

5. Avriel, M., Diewert, W.E., Schaible, S. and Zang, I. (1988) Generalized Concavity. Plenum Press, New York.

6. Bazaraa, M.S. and Shetty, C.M. (1976) Foundations of Optimization. Springer Verlag, Berlin.

7. Bazaraa, M.S., Sherali, H.D. and Shetty, C.M. (1993) Nonlinear Programming. John Wiley & Sons, New York.

8. Kemp, M.C. and Kimura, Y. (1978) Introduction to Mathematical Economics. Springer Verlag, New York.

9. Mangasarian, O.L. (1969) Nonlinear Programming. McGraw-Hill, New York.

10. Ponstein, J. (1967) Seven Kinds of Convexity. SIAM Review, 9, 115-119.

11. Simon, C.P. and Blume, L. (1994) Mathematics for Economists. W. W. Norton & Co., New York.

12. Cambini, A. and Martein, L. (2009) Generalized Convexity and Optimization. Springer, Berlin.

13. Crouzeix, J.-P. (2005) Criteria for Generalized Convexity and Generalized Monotonicity in the Differentiable Case. In: Hadjisavvas, N., Komlosi, S. and Schaible, S., Eds., Handbook of Generalized Convexity and Generalized Monotonicity, Springer, New York, 89-119.

3

CHAPTER

Efficient Numerical Methods for Solving Differential Algebraic Equations

Ampon Dhamacharoen

Department of Mathematics, Burapha University, Chonburi, Thailand

ABSTRACT

This research aims to solve Differential Algebraic Equation (DAE) problems in their original form, wherein both the differential and algebraic equations remain. The Newton or Newton-Broyden technique along with some integrators such as the Runge-Kutta method is coupled together to solve the problems. Experiments show that the method developed in this paper is efficient, as it demonstrates that implementation of the method is not difficult, and such method is able to provide approximate solutions with ease within some desired accuracy standards.

Keywords: Differential-Algebraic Equations, Newton-Broyden Method, Index-2 Hessenberg DAE

1. INTRODUCTION

A differential algebraic Equation (DAE) is an equation involving an unknown function and its derivatives. A DAE in its most general form is given by the following:

$$F\left(t, x(t), y(t), y'(t)\right) = 0, \quad t \in [a,b] \tag{1}$$

where R is the set of real number, $F : R \times R^m \times R^n \times R^n \rightarrow R^{m+n}$, $x(t) \in R^m$, $y(t) \in R^n$ and $y'(t) \in R^n$. In this form, the relationship between the variables and derivatives may be implicit. In some systems, the equations may be written in the explicit form of derivatives, as follows:

$$y'(t) = f\left(t, x(t), y(t)\right) \tag{2a}$$

$$g\left(t, x(t), y(t)\right) = 0 \tag{2b}$$

where $f : R \times R^m \times R^n \rightarrow R^n$, This set of equations is called a semi-explicit DAE system.

Differential algebraic equations arise in the mathematical modeling of a wide variety of problems found in engineering and science, such as multi-body and flexible body mechanics, electrical circuit design, optimal control, incompressible fluids, molecular dynamics, chemical kinetics and chemical process control [1] -[5] .

DAEs can be transformed into ODE problems via differentiation. The number of differentiations needed in the transforming process is called the differentiation index. This number can describe some characteristics of the problem. In general, the higher the index of a DAE, the more difficulties one can expect in its numerical solution [1] - [3] [6] .

Although DAEs can be transformed into an explicit ODE so that it can be solved using the methods of ODE, there are still many numerical methods that can solve DAE directly. In the DAE solvers software, the numerical approaches for the solution of DAEs can be divided roughly into two classes: a) direct discretizations of the given system; and b) methods which involve a reformulation (e.g. index reduction), combined with a discretization. Direct discretizations are easier to use, but are limited in their utilitisation to essentially index-1, index-2 Hessenberg, and index -3 Hessenberg DAE systems, while a reformulation may be costly, and it may also require more input from the user and involve more user intervention [3] [6] .

In this research, we will place emphasis on constructing an algorithm for solving a semi-explicit DAE. The approach employed is to firstly solve the system of algebraic equations, and subsequently solve the differential equations using the derived information. The Newton-Broyden method plays a key role in solving algebraic equations, since it performs almost as well as the Newton method, but requires less energy, and in addition, it also outperforms other methods of the same order of convergence [7] - [9] .

2. THE NEWTON-BROYDEN METHOD

The method, which was first proposed by Dhamacharoen in 2011 [7] , [8] , aims to solve the equation of the form:

$$\boldsymbol{F}\left(\boldsymbol{u}(\boldsymbol{x}),\boldsymbol{x}\right)=\boldsymbol{0} \tag{3}$$

The Newton scheme of this problem is:

$$x_{i+1} = x_i - \left[F_u\left(u\left(x_i\right), x_i\right) u'\left(x_i\right) + F_x\left(u\left(x_i\right), x_i\right) \right]^{-1} F\left(u\left(x_i\right), x_i\right), \quad i = 0,1,2,\cdots. \tag{4}$$

Replacing $u'(x_i)$ by D_i, gives

$$x_{i+1} = x_i - \left[F_u\left(u\left(x_i\right), x_i\right) D_i + F_x\left(u\left(x_i\right), x_i\right) \right]^{-1} F\left(u\left(x_i\right), x_i\right), \quad i = 0,1,2,\cdots. \tag{5}$$

with updating:

$$D_{i+1} = D_i - \frac{1}{b_i^{\mathrm{T}} b_i} \left(u\left(x_{i+1}\right) - u\left(x_i\right) - D_i b_i \right) b_i^{\mathrm{T}} \tag{6}$$

where $b_i = x_{i+1} - x_i$. Equation (6) is called the Broyden rank-1 update, and (5) with the update (6) is called the Newton-Broyden method. This method retains the good part of the Newton method, and replaces the difficult part of Newton's with Broyden's. With good initial guesses for $z_0$and D_0, the Newton-Broyden Scheme (5) will produce a sequence that converges to a solution of (3), with q-super linear order of convergence.

Although the order of convergence of the Newton-Broyden method is equal to that of the Broyden method, and less than that of the Newton method, in practice, the Newton-Broyden performs well in the sense that a good initial guess is easily found, and it reaches the solution in a reasonable number of ierations. As shown in [7] and [8] , for the same problem and using the same initial guess, the sequences from the Newton method and from the Newton-Broyden method reach the solution, while that from the Broyden does not. In addition, the Newton-Broyden method requires less amount of work in comparison with the Newton method.

Note that if function F is linear, then Broyden's update and Newton-Broyden's update coincide.

3. CONSTRUCTING THE METHOD

The initial value problems:

Assume that the DAE is expressed in the form (2a), (2b) (renumbering)

$$y'(t) = f(t, x(t), y(t)) \quad (7a)$$

$$g(t, x(t), y(t)) = 0 \quad (7b)$$

$$y(t_0) = y_0 \quad (7c)$$

in which the system of equations is to be solved for x(t) and y(t), where t Î [a, b]. In solving the system numerically, the conditions on the initial values of y must be imposed sufficiently for the system to have a unique solution. If (7b) can express x(t) in terms of t and y(t) explicitly, then the system becomes a pure ODE. Therefore, we consider the case when (7b) expresses x(t) implicitly.

Let the interval [a, b] be divided into n subintervals $\{a = t_0, t_1, \cdots, t_n = b\}$. In each interval $[t_{i-1}, t_i]$, we will solve (7a) numerically for $y(t_i)$. In an initial value problem, the value $y(t_0)$ is specified, and the value $x(t_0)$ can subsequently be solved from (7b). In order to use the Runge-Kutta method, in each interval $[t_{i-1},t_i]$ the value x_1, x_2 and x_3, which are needed for computing F_2, F_3 and F_4 respectively, can each be solved from (7b) using the values of $y(t_{i-1})+F_1/2$ in F_2, $y(t_{i-1})+F_2/2$ in F_3 and $y(t_{i-1})+F_3/2$ in F_4. Once the value $y(t_i)$ is computed, $x(t_i)$ can then be solved from (7b). Advance to the next interval, and repeat this procedure until we reach the last subinterval.

Suppose the term x(t) is missing from the Equation (7b). Therefore, the system becomes:

$$y'(t)=f(t,x(t),y(t)) \tag{8a}$$

$$g(t,y(t))=0 \tag{8b}$$

$$y(t_0)=y_0 \tag{8c}$$

in which the system is called index-2 Hessenberg. In solving the differential Equation (8a), the variable x(t) are treated as unknowns. In each interval $[t_{i-1},t_i]$, the value of $x(t_{i-1})$ is given as a guess, and (8a) is then solved

numerically for $y(t_i)$. Check the condition (8b) $g(t_i,y(t_i))=0$. Update the value of $x(t_0)$ and iterate the process until the condition (8b) is met. Subsequently, advance to the next interval using the last value of $x(t_{i-1})$ as the guessing value for $x(t_i)$. Repeat this procedure until we complete the last subinterval.

In solving the differential equations, we may use the 4th order Runge-Kutta method, or the 4thorder Taylor method, since their local error is $O\left(h^5\right)$ and the total error is $O\left(h^4\right)$, which is acceptable and also easy to implement. In solving the algebraic equations associated in the problem, the Newton method is used in (7b), and the Newton-Broyden method is used in (8b).

As per the procedure described above, the formulation will be as follows:

Partition the interval [a, b] into n subintervals. $h=\frac{b-a}{n}$,

and $t_0=a,\ t_i=t_{i-1}+h,\ i=1,2,\cdots,n.$

Problem (7a), (7b):

Define the function F: $F\left(x\left(t_{i-1}\right)\right)=g\left(t_i,x\left(t_i\right),y\left(t_i\right)\right)$, $i=1,2,\cdots,n$.

The problem become (3)

$$F\left(x\left(t_{i-1}\right)\right)=0$$

which can be solved using the Newton method.

P1: Newton Method.

Prescribe a small positive real number e.

Guess the value z.

A: Compute F(z).

Check if ||F(z)|| < e

If so, proceed to B. If not, carry out the next step.

Compute F'(z).

Solve the system F'(z)b = −F(z), for b.

Set z = z + b.

Go to A.

B: End.

The process for solving the DAE will be as follows:

Algorithm A:

Initial step:

1) Set the given initial condition $y(t_0) = y_0$.

2) Solve for $x(t_0)$, using P1: Set $x(t_0) = z$.

Main step:

For $i = 1, 2, \cdots, n$; process the following steps:

Step 1. Solve the initial value problem (7a) 1 step to obtain $y(t_i)$, by proceeding using the following steps:

1) Compute: $F_1 = hf\left(t_{i-1}, x(t_{i-1}), y(t_{i-1})\right)$, $y_1 = y(t_{i-1}) + F_1/2$,

Solve $g(t_{i-1} + h/2, x_1, y_1) = \mathbf{0}$ for x_1. (Using P1).

2) Compute: $F_2 = hf\left(t_{i-1} + h/2, x_1, y_1\right)$

$y_2 = y(t_{i-1}) + F_2/2$, Solve $g(t_{i-1} + h/2, x_2, y_2) = 0$ for x_2. (Using P1).

3) Compute: $F_3 = hf(t_{i-1} + h/2, x_2, y_2)$

$y_3 = y(t_{i-1}) + F_3/2$, Solve $g(t_{i-1} + h, x_3, y_3) = 0$ for x_3. (Using P1).

4) Compute: $F_4 = hf(t_{i-1} + h, x_3, y_3)$

5) $y(t_i) = y(t_{i-1}) + (F_1 + 2(F_2 + F_3) + F_4)/6$

Step 2. Solve (7b) for $x(t_i)$, using P1: Set $x(t_i) = z$.

Proceed to the next interval.

Problem (8a), (8b):

Define the function F: $F(x(t_{i-1}), y(t_{i-1})) = g(t_i, y(t_i))$, $i = 1, 2, \cdots, n$.

The problem become (3)

$$F(x(t_{i-1}), y(t_{i-1})) = 0$$

which can be solved using the Newton-Broyden method.

P2: One Step Runge-Kutta Method

Given the value $t_{i-1}, x(t_{i-1}), y(t_{i-1})$ and b.

Compute: $F_1 = hf\left(t_{i-1}, x\left(t_{i-1}\right), y\left(t_{i-1}\right)\right)$, $y_1 = y\left(t_{i-1}\right) + F_1/2$, $x_1 = x\left(t_{i-1}\right) + b/2.$

Compute: $F_2 = hf\left(t_{i-1} + h/2, x_1, y_1\right)$, $y_2 = y\left(t_{i-1}\right) + F_2/2$, $x_2 = x\left(t_{i-1}\right) + b/2.$

Compute: $F_3 = hf\left(t_{i-1} + h/2, x_2, y_2\right)$, $y_3 = y\left(t_{i-1}\right) + F_3$, $x_3 = x\left(t_{i-1}\right) + b.$

Compute: $F_4 = hf\left(t_{i-1} + h, x_3, y_3\right).$

Then $y\left(t_i\right) = y\left(t_{i-1}\right) + \left(F_1 + 2\left(F_2 + F_3\right) + F_4\right)/6.$

The process for solving the DAE will be as follows:

Algorithm B:

Initial step:

1) Set the given initial condition $y\left(t_0\right) = y_0$.

2) Guess the initial condition $x\left(t_0\right) = z$. $x\left(t_0\right) = b$.

3) Guess the initial matrix D (may be = I), Guess the value b.

4) Solve the initial value problem (8a) 1 step, to obtain $y\left(t_1\right)$, using P2.

5) Compute the value $F\left(x\left(t_0\right), y\left(t_0\right)\right) = g\left(t_1, y\left(t_1\right)\right).$

Main step:

For $i = 1, 2, \cdots, n$; process the following steps:

Step 1. Check the condition $\left\| g\left(t_i, y\left(t_i\right)\right) \right\| < \varepsilon$ where e is a prescribed small number.

If it passes, proceed to the next interval (next i).

If it fails, carry out step 2.

Step 2.

1) Set $u = y\left(t_i\right)$.

2) Compute the matrix $A = F_x\left(x\left(t_{i-1}\right), y\left(t_i\right)\right) + F_y\left(x\left(t_{i-1}\right), y\left(t_i\right)\right) D$.

3) Solve $Ab = -g\left(t_i, y\left(t_i\right)\right)$ for b.

4) Set $x\left(t_{i-1}\right) = x\left(t_{i-1}\right) + b$.

5) Solve the initial value problem (8a) 1 step (using P2) to obtain $y\left(t_i\right)$.

6) Compute the new value $g\left(t_i, y\left(t_i\right)\right)$.

7) Update the matrix $D = D + \frac{1}{b^{\mathrm{T}} b}\left(y\left(t_i\right) - u - Db\right) b^{\mathrm{T}}$.

8) Go to step 1.

End.

4. EXPERIMENTS

Three examples are used to illustrate the method. The first problem is an index-1 Hessenberg DAE system, with nonlinear differential equations and a nonlinear algebraic equation, as follows:

$$x''(t) = -(3t+1)y(t) - x(t)(4z(t)+1), \quad 0 \le t \le 1$$

$$y''(t) = 4\cos(z(t)) - y(t)(4z(t)+1)$$

initial conditions,

$$x(0)=0, \; y(0)=0,$$

$$x'(0)=1, \; y'(0)=2$$

algebraic equation

$$4x(t)\cos(z(t)) + ty^2(t) = 4(z(t)-t^2)$$

Using Algorithm A, this problem is solved and has a nice result with a small error as compared to the exact solution

$$z(t) = t(t+1)$$

$$x(t) = t\cos(z(t)),$$

$$y(t) = \sin(z(t)).$$

Some results are illustrated in Table 1. (xs, ys and zs are values from the exact solution).

The second problem is a classical example of the DAE problem which is "The pendulum problem", as expressed in the xy co-ordinate plane as follows:

$$\left.\begin{aligned} x''(t) &= -l(t)x(t), \quad t \geq 0 \\ y''(t) &= -l(t)y(t) - g \\ x(t)^2 + y(t)^2 &= L^2 \end{aligned}\right\} \tag{8}$$

where g = 9.8. This problem is an index 2 semi-explicit DAE problem. [3] , [10] . Proceeding to solve the problem numerically, we let L = 1, and impose the initial conditions x(0) = 1, y(0) = 0. Using the new variables $y_1 = x,\ y_2 = x',\ y_3 = y$ and $y_4 = y'$, we have a system of differential algebraic equations as follows:

Table 1. Resulting values for x(t), y(t) and z(t) from Algorithm A, and the errors.

i	t	x	y	z	$x-xs$	$y-ys$	$z-zs$
0	0.000000	0.000000	0.000000	0.000000	0.000000	0.000000	0.000000
5	0.083333	0.082994	0.180310	0.090278	0.000000	0.000000	0.000000
10	0.1666667	0.1635259	0.3864430	0.1944444	−0.000000	0.0000000	−0.000000
15	0.2500000	0.2378920	0.6148770	0.3125000	−0.000000	0.0000000	−0.000000
20	0.3333333	0.3009499	0.8599127	0.4444444	−0.000000	0.0000000	−0.000000
25	0.4166667	0.3461609	1.1131836	0.5902778	−0.000000	0.0000000	−0.000000
30	0.5000000	0.3658444	1.3632775	0.7500000	−0.000000	0.0000000	−0.000000
35	0.5833333	0.3517169	1.5955682	0.9236111	−0.000000	0.0000000	−0.000000
40	0.6666667	0.2957773	1.7923844	1.1111111	−0.000000	0.0000000	−0.000000
45	0.7500000	0.1915753	1.9336531	1.3125000	−0.000000	0.0000000	0.0000000
50	0.8333333	0.0358377	1.9981497	1.5277778	0.0000000	0.0000000	0.0000000
55	0.9166667	−0.169652	1.9654489	1.7569445	0.0000001	0.0000001	0.0000001
60	1.0000000	−0.416147	1.8185950	2.0000002	0.0000002	0.0000003	0.0000002

$$\left.\begin{aligned} y_1' &= y_2 \\ y_2' &= -\lambda y_1 \\ y_3' &= y_4 \\ y_4' &= -\lambda y_3 - g \end{aligned}\right\}$$

initial conditions:

$$\begin{aligned} y_1(0) &= 1 \\ y_2(0) &= 0 \\ y_3(0) &= 0 \\ y_4(0) &= 0 \end{aligned} \qquad (9a)$$

algebraic equation:

$$y_1^2 + y_3^2 - 1 = 0 \qquad (9b)$$

If we let $x = \sin\theta$, $y = -\cos\theta$, substitute in the problem (8) and manipulate some derivatives and algebra, we will obtain the equivalent initial value problem of an ordinary differential equation:

$$\left.\begin{aligned} &\theta''(t) = -g\sin(\theta(t)), \quad t \geq 0 \\ &\theta(0) = \frac{\pi}{2} \end{aligned}\right\} \qquad (10)$$

From (9b) if we differentiate the equation $y_1(t)^2 + y_3(t)^2 = 1$ twice, we will obtain the equation:

$$\lambda(t) = y_2(t)^2 + y_4(t)^2 - gy_2(t). \tag{11}$$

The system (9a) with (11) becomes a pure ODE, since the function l(t) is expressed in the terms $y_2(t)$ and $y_4(t)$ explicitly.

In this experiment, we solve (9a), (9b) using Algorithm B, and provide some of the results in Table 2. System (10) and (9a) with (11) are also solved, and they provide the same results, wherein their values are used to com pare to the results from Algorithm B, as shown in Table 2. The value xs, ys and ls are from problem (10). Note that the value of l(t)'s are at the mid-points t + h/2.

Table 2. Result values for x (t), y (t) and l (t) from Algorithm B, and the errors.

i	t	x	y	$\lambda(t)$	$x-xs$	$y-ys$	$\lambda-\lambda s$
0	0.000000	1.000000	0.000000	0.0066694	0.000000	0.000000	−0.01334
5	0.083333	0.999421	−0.034020	1.2067764	0.000000	0.000000	−0.01327
10	0.166667	0.990763	−0.135608	4.3889120	0.000000	0.000000	−0.01245
15	0.250000	0.953758	−0.300577	9.4089741	−0.000000	−0.000000	−0.00924
20	0.333333	0.858149	−0.513401	15.762372	−0.000000	−0.000000	−0.001840
25	0.416667	0.674223	−0.738528	22.336944	0.000000	0.000000	0.009746
30	0.500000	0.392046	−0.919946	27.443990	0.000000	0.000000	0.021502
35	0.583333	0.039516	−0.999219	29.406564	0.000001	0.000000	0.026667
40	0.666667	−0.320679	−0.947188	27.497515	0.000001	−0.000000	0.021650
45	0.750000	−0.621722	−0.783238	22.424766	−0.000001	0.000001	0.009949
50	0.833333	−0.826959	−0.562262	15.858262	0.000001	−0.000000	−0.001681
55	0.916667	−0.939329	−0.343017	9.4922857	0.000000	−0.000000	−0.009152
60	1.0000000	−0.986140	−0.165918	4.4483962	0.000000	−0.000000	−0.012414

The third problem is an index-2 Hessenberg DAE system with nonlinear differential equations and a nonlinear algebraic equation, as follows:

$$x''(t) = x(t)(4z(t)-1) + 2(1-3t)y(t), \quad 0 \le t \le 1$$

$$y''(t) = 2\sin(z(t)) + y(t)(4z(t)-1)$$

initial conditions,

$$x(0) = 0, \quad y(0) = 1,$$

$$x'(0) = 0, \quad y'(0) = 0$$

algebraic equation

$$x^2(t) + t^2 y^2(t) = t^2$$

Note that this problem is similar to the first problem, but there is no variable z(t) in the algebraic equation.

Using Algorithm B, this problem is solved, and provides a nice result as shown in Table 3. The exact solution is as follows:

$$z(t) = t(1-t)$$

$$x(t) = t\sin(z(t)),$$

$$y(t) = \cos(z(t)).$$

Algorithm B gives the results for x(t) and y(t), with a small error. However, the solution for z(t) has some errors less than 0.005. Some of the results are illustrated in Table 3. Note that the value of z(t)'s are at the mid- points t + h/2.

Table 3. Result values for x (t), y (t) and z (t) from Algorithm A, and the errors.

i	t	x	y	z	$x-xs$	$y-ys$	$z-zs$
0	0.0000000	0.0000000	1.0000000	.0055058	0.0000000	0.0000000	−0.0027581
5	0.0833333	0.0063659	0.9970838	0.0860924	0.0000000	−0.0000000	0.0028285
10	0.1666667	0.0230738	0.9903704	0.1416003	0.0000000	−0.0000000	−0.0027746
15	0.2500000	0.0466008	0.9824733	0.1944127	0.0000000	−0.0000000	0.0028155
20	0.3333333	0.0734659	0.9754101	0.2221708	0.0000000	−0.0000000	−0.0027597
25	0.4166667	0.1002790	0.9706071	0.2471710	0.0000000	−0.0000000	0.0027960
30	0.5000000	0.1237020	0.9689124	0.2471938	0.0000000	−0.0000000	−0.0027367
35	0.5833333	0.1403905	0.9706071	0.2443677	0.0000000	−0.0000000	0.0027705
40	0.6666667	0.1469318	0.9754101	0.2166649	0.0000000	−0.0000000	−0.0027101
45	0.7500000	0.1398025	0.9824733	0.1860087	0.0000000	−0.0000000	0.0027448
50	0.8333333	0.1153690	0.9903704	0.1305767	0.0000000	−0.0000000	−0.0026872
55	0.9166667	0.0699551	0.9970838	0.0721021	0.0000000	−0.0000000	0.0027271
60	1.0000000	0.0000001	1.0000000	−0.0110809	0.0000000	0.0000000	−0.0026781

5. CONCLUSION

The methods are constructed with the objective of solving DAE systems in their original forms. Both algorithms use the Runge-Kutta method as the integrator, and couple this with a method to solve the algebraic systems associated in the problem. The Newton method is used in Algorithm A for index-1 DAE, while in Algorithm B the Newton-Broyden method is needed for an index-2 DAE system. The methods can give approximate solutions for the problem very well, with only small errors. Experiments have also shown that high index DAEs are harder to solve than lower index DAEs.

ACKNOWLEDGEMENTS

The author would like to thank the referees for their comments. This work was financially supported by the Research Grant of

Burapha University through National Research Council of Thailand (Grant No. 35/2556).

REFERENCES

1. Ascher, U.M. and Petzold, L.R. (1998) Computer Method for Ordinary Differential Equations and Differential-Algebraic Equations. Society for Industrial and Applied Mathematics (SIAM), Philadelphia.

2. Brenan, K.E., Campbell, S.L. and Petzold, L.R. (1996) Numerical Solution of Initial-Value Problems in Differential-Algebraic Equations, Revised and Corrected Reprint of 1989 Original, with an Additional Chapter and Additional References. Classic in Applied Mathematics, 14. SIAM, Philadelphia.

3. Campbell, S.L., et al. (2008) Differential-Algebraic Equations. Scholarpedia, 3, 2849.

4. Rabier, P.J. and Rheinboldt, W.C. (2002) Theoretical and Numerical Analysis of Differential-Algebraic Equations. Hand Book of Numerical Analysis, 8, 183-540.

5. Riaza, R. (2008) Differential-Algebraic System: Analytical Espects and Circuit Applications. World Scientific, Singapore City.

6. Hairer, E. and Wanner, G. (1998) Solving Ordinary Differential Equation. II. Stiff and Differential-Algebraic Problems, 2nd Edition, Springer Series in Computational Mathematics, 14. Springer-Verlag, Berlin.

7. Chompuvised, K. and Dhamacharoen, A. (2011) Solving Boundary Value Problems of Ordinary Differential Equations

with Non-Separated Boundary Conditions. Applied Mathematics and Computation, 217, 10355-10360.

8. Dhamacharoen, A. (2014) An Efficient Hybrid Method for Solving Systems of Nonlinear Equations. Journal of Computational and Applied Mathematics, 263, 59-68.

9. Dhamacharoen, A. and Chompuvised, K. (2013) An Efficient Method For Solving Multipoint Equation Boundary Value Problems. World Academy of Science, Engineering and Technology, 7, 329-333.

10. Differential-Algebraic Equation, Wikipedia, the Free Encyclopedia.

4

CHAPTER

Numerical Solution of Second-Order Linear Fredholm Integro-Differetial Equations by Trigonometric Scaling Functions

Hamid Safdari, Yones Esmaeelzade Aghdam

Department of Mathematics, Shahid Rajaee Teacher Training University, Tehran, Iran

ABSTRACT

The main aim of this paper is to apply the Hermite trigonometric scaling function on [0, 2p] which is constructed for Hermite interpolation for the linear Fredholm integro-differential equation of second order. This equation is usually difficult to solve analytically. Our approach consists of reducing the problem to a set of algebraic linear equations by expanding the approximate solution. Some numerical example is included to demonstrate the validity and applicability of the presented technique, the method produces very accurate results, and a comparison is made with exiting re- sults. An estimation of error bound for this method is presented.

Keywords: Numerical Technique, Fredholm Integro-Differential Equations, Hermite Trigonometric Wavelets, Operational Matrix, Error Estimates

1. INTRODUCTION

In this paper we solve the Fredholm Linear Integro-Differential Equations as

$$\mu_0(x)+\mu_1(x)u'(x)+\mu_2(x)u''(x)=g(x)+\int_0^1 k(x,t)u(t)dt,\quad 0\le x\le 1,\ u(0)=u_0,\ u(1)=u_1 \tag{1}$$

where $\mu_i(x)$, $f(x)$, and $K_l(x,t)$ are given functions that have suitable derivatives, and u_0 and u_1 are given real constans. In most situations, it is difficult to obtain exact solution of the above integration. Hence various approximation method have been proposed and studied. The purpose of the present paper is to develop a trigonometric Hermite wavelet approximation for the computing of the problem [1] .

Systems of integro-differential equations have a major role in the fields of science, physical phenomena, and engineering, such as nano-hydrodynamics, glass-forming process, dropwise condensation, wind ripple in the de- sert, and modeling the competition between tumor cells and the immune system. The concept of a system of integro-differential equations has motivated a huge amount of research work in recent years. Alot of attention has been devoted to the study of differential-difference equations, e.g. equations containing shifts of the un- known function and its derivates, and also integro-differential-difference equations. For instance, see [2] [3] . There are several numerical methods for solving system of linear integro-differential equations, for example, the rationalized Haar functions method [4] , Galerkin methods with hybrid functions [5] , the spline approximation method [6] , the Chebyshev polynomial method [7] , the spectral method [8] , the CAS wavelet method [9] , Ruge-Kutta methods [10] , the Adomian decomposition methods [11] , and the interested reader can see [12] [13] for more published research works in the subject.

Our approach consists of reducing the problem to a set of linear equations by trigonometric scaling functions which is constructed for Hermite interpolation. A difficulty of using wavelet for the representation of integral operators is that quadrature leads to potentially high cost with sparse matrix. This fact particularly encourages us in efforts to devote to some appropriate wavelet bases to simplify the computation expense of the reoresentation matrix, which is importent to improve the wavelet method. Recently, the trigonometric interpolant wavelet has arisen in the approximation of operators [14] - [16] . Quack [17] has constructed a multiresolution analysis (MRA). Chen [18] [19] presented the feasibility of trigonometric wavelet numerical methods for stokes problem and Hadamard integral equation.

The organization of the rest of this paper is as follows: Section (0) describe the trigonometric scaling function on $[0,2\pi]$, and construct the operational matrix of derivative for these function. Section (0) summarizes the application of trigonometric scaling functions to the solution of Problem (1). Thus, a set of linear equations is formed and a solution of the considered problem is introduced. In Section (0), we report our computational results and demonstrate the accuracy of the proposed numerical schemes by presenting numerical examples. Note that we have computed the numerical results by MATLAB programming.

2. INTERPOLATORY HERMITE TRIGONOMETRIC WAVELETS

In this section, we will give a brief introduction of Quak's work on the construction of Hermite interpolatory trigonometric wavelets and their basic properties. More details can be found in (see [17]).

For all $j \in \mathbb{N} = \mathbb{N} \bigcup \{0\}$, two scaling functions $\varphi_{j,0}^{0}(x)$ and $\varphi_{j,0}^{1}(x)$ are defined as

$$\varphi_{j,0}^{0}(x) = \frac{1}{2^{2j+1}} \sum_{l=0}^{2^{j+1}-1} D_l(x) \tag{2}$$

$$\varphi_{j,0}^{1}(x) = \frac{1}{2^{2j+1}} \left(\tilde{D}_{2^{j+1}}(x) + \frac{1}{2} \sin\left(2^{j+1} x\right) \right) \tag{3}$$

where the Dirichlet kernel $D_l(x)$ and its conjugate kernel $\tilde{D}_l(x)$ are defined as

$$D_l(x) = \frac{1}{2} + \sum_{k=1}^{l} \cos kx = \begin{cases} \dfrac{\sin\left(l + \frac{1}{2}\right)x}{2 \sin \frac{x}{2}} & x \notin 2\pi\mathbb{Z} \\ l + \dfrac{1}{2} & x \in 2\pi\mathbb{Z} \end{cases}$$

$$\tilde{D}_l(x) = \sum_{k=1}^{l} \sin kx = \begin{cases} \dfrac{\cos \frac{x}{2} - \cos\left(l + \frac{1}{2}\right)}{2 \sin \frac{x}{2}} & x \notin 2\pi\mathbb{Z} \\ 0 & x \in 2\pi\mathbb{Z} \end{cases}$$

Obviously, $D_l(x), \tilde{D}_l(x) \in T_l$, where T_l is the linear space of trigonometric polynomials with degree not

exceeding l. The equally spaced nodes on the interval $[0, 2\pi)$

with a dyadic step are denoted by $t_{j,l} = \frac{l\pi}{2^j}$, for

any $j \in \mathbb{N}_0$ and $l = 0, 1, \cdots, 2^{j+1} - 1$, where $\mathbb{N}_0$ is the set of all non-negative integers. Let

$\varphi^s_{j,l}(x) = \varphi^s_{j,0}(x - t_{j,l})$, for $s = 0, 1$, and $l = 0, 1, \cdots, 2^{j+1} - 1$.

Lemma 1 (See [17] .) For $j \in \mathbb{N}_0$, we have

$$\varphi^0_{j,0}(x) = \begin{cases} \dfrac{1}{2^{2j+2}} \dfrac{\sin^2(2^j x)}{\sin^2\left(\dfrac{x}{2}\right)} & x \notin 2\pi\mathbb{Z} \\ 1 & x \in 2\pi\mathbb{Z} \end{cases}$$

$$\varphi^1_{j,0}(x) = \begin{cases} \dfrac{1}{2^{2j+2}} \left(1 - \cos\left(2^{j+1} x\right)\right) \cot\left(\dfrac{x}{2}\right) & x \notin 2\pi\mathbb{Z} \\ 0 & x \in 2\pi\mathbb{Z} \end{cases}$$

and their derivations are given by

$$\left(\varphi_{j,0}^{0}(x)\right)' = \begin{cases} \dfrac{1}{2^{j+2}} \dfrac{\sin\left(2^{j+1}x\right)}{\sin^2\left(\dfrac{x}{2}\right)} - \dfrac{1}{2^{2j+2}} \dfrac{\sin^2\left(2^{j}x\right)\cot\left(\dfrac{x}{2}\right)}{\sin\left(\dfrac{x}{2}\right)} & x \notin 2\pi\mathbb{Z} \\ 0 & x \in 2\pi\mathbb{Z} \end{cases}$$

$$\left(\varphi_{j,0}^{1}(x)\right)' = \begin{cases} \dfrac{1}{2^{2j+3}} \dfrac{\cos\left(2^{j+1}x\right)-1}{\sin^2\left(\dfrac{x}{2}\right)} + \dfrac{1}{2^{j+1}} \sin\left(2^{j+1}x\right)\cot\left(\dfrac{x}{2}\right) & x \notin 2\pi\mathbb{Z} \\ 1 & x \in 2\pi\mathbb{Z} \end{cases}$$

Theorem 1 (Interpolatory properties of the scaling functions). (See [17]) For $j \in \mathbb{N}_0$, the following inter- polatory properties hold for each $l,k = 0,1,\cdots,2^{j+1}-1$

$$\varphi_{j,l}^{0}\left(x_{j,k}\right) = \delta_{k,l}, \quad \left(\varphi_{j,l}^{0}\left(x_{j,k}\right)\right)' = 0,$$
$$\varphi_{j,l}^{1}\left(x_{j,k}\right) = 0, \quad \left(\varphi_{j,l}^{1}\left(x_{j,k}\right)\right)' = \delta_{k,l}. \tag{4}$$

From above we can take wavelet functions $\varphi_{j,l}^{0}(x), \varphi_{j,l}^{1}(x), l = 0,1,\cdots,2^{j+1}-1$ as scaling functions. Now, we can define the scaling function spaces V_j. Then we have

Definition 2 (Scaling functions space). For all $j \in \mathbb{N}_0$ define the wave space V_j as follows

$$V_j = span\left\{\varphi_{j,n}^{0}(t), \varphi_{j,n}^{1}(t), n = 0,1,\cdots,2^{j+1}-1\right\}$$

As a first step of studying the spaces V_j, the following result identifies the trigonometric polynomials which from alternative bases of these spaces.

Now a Hermite-type project operator can be introduced by means of the scaling functions. For all $j \in \mathbb{N}_0$ the Hermite projection operators L_j mapping any real-valued differentiable 2π-periodic function f into the space V_j is defined as

$$f(x)L_j f(x) \simeq \sum_{k=0}^{2^{j+1}-1}\left[a_k \varphi_{j,k}^0(x) + b_k \varphi_{j,k}^1(x)\right] = C^{\mathrm{T}}\Phi \tag{5}$$

where $a_k = f(t_{j,k})$, $b_k = f'(t_{j,k})$, C, and Φ are vectors with dimension $2^{j+2} \times 1$. The following properties of the operators L_j are therefore obvious:

$L_j \in T_{2^{j+1}}$

$$L_j f(x_{j,k}) = f(x_{j,k}) \text{ and } (L_j f)'(x_{j,k}) = f'(x_{j,k}), k \in \mathbb{Z}$$

$L_j f = f$ for all $f \in V_j$.

Theorem 3 Let $f(x) \in L_{2\pi}^2$, and its trigonometric wavelet approximation is $L_J f$, then we have

$$\|f(x) - L_J f(x)\|_{L_{2\pi}^2} \le C 2^{-2(J+1)}$$

where C is a positive constant value.

Proof. See [17] .

Lemma 2 (The operational matrix of scaling function derivative). (See [20]) The differentiation of vector Φ in 5 can be expressed as [20]

$$\Phi' = D_\phi \Phi$$

where D_ϕ is $2^{J+2} \times 2^{J+2}$ operational matrix of derivative for trigonometric scaling function. Suppose

$$\left(\phi_{j,m}^{s}(x)\right)' = \sum_{k=0}^{2^{j+1}-1}\left[a_{k,m}^{s}\varphi_{j,k}^{0}(x)+b_{k,m}^{s}\varphi_{j,k}^{1}(x)\right] \tag{6}$$

where $m = 0,1,\cdots,2^{j+1}-1$ and $s = 0,1$. So the matrix D_ϕ can be respresented as a block matrix as

$$D = \begin{bmatrix} A^0 & B^0 \\ A^1 & B^1 \end{bmatrix}$$

where A^s and B^s are $2^{j+1} \times 2^{j+1}$ matrices. The entries of matrices A^s and B^s may be find by using

$$A^s = \left(a_{k,m}^{s}\right) = \left(\phi_{j,k}^{s}\right)'\left(x_{j,m}\right), \quad B^s = \left(b_{k,m}^{s}\right) = \left(\phi_{j,k}^{s}\right)''\left(x_{j,m}\right)$$

where A^0 is a $2^{j+1} \times 2^{j+1}$ zero matrix, A^1 is a $2^{j+1} \times 2^{j+1}$ identity matrix. Using $x_{j,m} = \dfrac{m\pi}{2^j}$ we get

$$\phi_{j,k}^{s}\left(x_{j,m}\right)=\phi_{j,0}^{s}\left(\frac{(m-k)\pi}{2^{j}}\right)=\phi_{j,0}^{s}\left(x_{j,m-k}\right) \tag{7}$$

Using 7 and $B^{s}=\left(b_{k,m}^{s}\right)=\left(\phi_{j,k}^{s}\left(x_{j,m}\right)\right)''=\left(\phi_{j,0}^{s}\left(x_{j,m-k}\right)\right)'' \quad s=0,1$ we get

$$B^{0}=\left(b_{k,m}^{0}\right)=\left(\phi_{j,0}^{0}\right)''\left(x_{j,m-k}\right)=\begin{cases}\dfrac{1}{2}\dfrac{\cos\left((m-k)2\pi\right)}{\sin^{2}\left(\dfrac{m-k}{2^{j+1}}\pi\right)} & k\neq m\\[2ex] \dfrac{-1}{2^{2j+1}}\displaystyle\sum_{p=0}^{2^{j+1}-1}\sum_{k=1}^{p}\frac{p(p+1)(2p+1)}{6} & k=m\end{cases} \tag{8}$$

and

$$B^{1}=\left(b_{k,m}^{1}\right)=\left(\phi_{j,0}^{1}\right)''\left(x_{j,m-k}\right)\begin{cases}-\cot\left(\dfrac{m-k}{2^{j+1}}\pi\right) & k\neq m\\[2ex] 0 & k=m\end{cases} \tag{9}$$

for $k,m=0,1,\cdots,2^{j+1}-1$.

3. PROCEDURE SOLUTION USING THE TRIGONOMETRIC SCALING FUNCTION

In this section, we first give the computational schemes for Equation (1) with the Newton-Cotes formulas. For either one of these rules, we can make a more accurate approximation by breaking up the interval $[0,1]$ into some number N of subintervals. This is called a composite rule, extended rule, or iterated rule. For example, the composite trapezoidal rule for the discretization form of (1) can be stated as

$$\sum_{i=0}^{2}\mu_i(x)u^{(i)}(x)=g(x)+\frac{1}{N}\left(\frac{k(x,0)u(0)}{2}+\sum_{k=1}^{N-1}\left(k\left(x,\frac{k}{N}\right)u\left(\frac{k}{N}\right)\right)+\frac{k(x,1)u(1)}{2}\right),$$
$$0\le x\le 1,\ u(0)=u_0,\ u(1)=u_1 \tag{10}$$

where the subintervals have the form $[kh,(k+1)h]$, with $h=\frac{1}{N}$ and $k=0,1,\cdots,N-1$. By introducing a basis $\{\phi_{j,k}\}$ for the subspace V_J, the coefficients vector $u(x)$ of the discrete solution is defined by

$$GL_j u(x)=\sum_{k=0}^{2^{j+1}-1}\left[a_k\varphi^0_{j,k}(x)+b_k\varphi^1_{j,k}(x)\right]=C^{\mathrm{T}}\Phi \tag{11}$$

where C is $2^{J+2}\times 1$ unknown vector defined similar to (5). By substituting $L_j u(x)$ and using Lemma (2) in (1) we have a linear system. Now for determining unknown coefficients vector C or a_k and b_k, we choose collo- cation method with choosing collocation points as

$$t_\tau=a+\frac{\tau(b-a)}{2^{J+2}}=\frac{\tau\pi}{2^{J+1}};\quad \tau=0,1,\cdots,2^{J+2}-1 \tag{12}$$

Thus we have

$$\sum_{i=1}^{2}\mu_i(t_\tau)L_ju^{(i)}(t_\tau)-\frac{1}{N}\sum_{k=1}^{N-1}\left(k\left(t_\tau,\frac{k}{N}\right)L_ju\left(\frac{k}{N}\right)\right)$$
$$=\frac{1}{2N}\left(k(t_\tau,0)L_ju_0+k(t_\tau,1)L_ju_1\right)+g(t_\tau)-\mu_0(t_\tau), \qquad (13)$$

By using Lemma 2 and after summarizing Equation (13) can be rewritten as the matrix form $ZX=F$ where $k=0,1,\cdots,2^{J+1}-1$, $i,j=1,\cdots,2^{J+1}$, and $\tau=0,1,\cdots,2^{J+2}-1$.

Now, let us calculate the entries $z_{m,s}^{m,s}, m,s=1,2$ and $f(t_\tau)$ in the system matrix.

$$Z_{2^{J+2}\times 2^{J+2}}=\begin{bmatrix} Z_{1,1} & Z_{1,2} \\ Z_{2,1} & Z_{2,2}\end{bmatrix}$$

$$Z_{1,1}=\{\mu_1\}B^0-\left\{\frac{1}{N}\sum_{k=1}^{N-1}\left(k\left(t_\tau,\frac{k}{N}\right)L_ju\left(\frac{k}{N}\right)\right)\right\}I=\left(z_{i,j}^{1,1}\right)_{2^{J+1}\times 2^{J+1}}$$

$$Z_{2,1}=\{\mu_1\}B^1+\left\{\mu_0-\frac{1}{N}\sum_{k=1}^{N-1}\left(k\left(t_\tau,\frac{k}{N}\right)L_ju\left(\frac{k}{N}\right)\right)\right\}I=\left(z_{i,j}^{2,1}\right)_{2^{J+1}\times 2^{J+1}}$$

$$Z_{1,2}=B^0\left\{\left(\mu_0+\mu_1B^1\right)\right\}-\left\{\frac{1}{N}\sum_{k=1}^{N-1}\left(k\left(t_\tau,\frac{k}{N}\right)L_ju\left(\frac{k}{N}\right)\right)\right\}I=\left(z_{i,j}^{1,2}\right)_{2^{J+1}\times 2^{J+1}}$$

$$Z_{2,2}=\{\mu_0\}B^1+\mu_1\left(B^0+B^{12}\right)-\left\{\frac{1}{N}\sum_{k=1}^{N-1}\left(k\left(t_\tau,\frac{k}{N}\right)L_ju\left(\frac{k}{N}\right)\right)\right\}I=\left(z_{i,j}^{2,2}\right)_{2^{J+1}\times 2^{J+1}}$$

$$X_{2^{J+2}\times 1}=\begin{bmatrix} X_1 \\ X_2 \end{bmatrix} \quad X_1=(a_k)_{2^{J+1}\times 1} \quad X_2=(b_k)_{2^{J+1}\times 1}$$

$$F=(f(t_\tau))_{2^{J+2}\times 1} \quad f(t_\tau)=\frac{1}{2N}(k(t_\tau,0)L_J u_0+k(t_\tau,1)L_J u_1)+g(t_\tau)-\mu_0(t_\tau)$$

where the matrix A^0, A^1, B^0, and B^1 defined in Lemma (2), and I is a $2^{j+1}\times 2^{j+1}$ identity matrix. So the unknown function $u_J(x)$ can be found. Note that we find these function by MATLAB.

4. NUMERICAL EXAMPLE

To support our theoretical discussion, we applied the method presented in this paper to several examples. The main objective here is to solve these two examples using the trigonometric scaling function and compare our results with exact solution.

Example 4 Consider the second-order the Fredholm Linear Integro-Differential Equation

$$y''(x)+4xy'=-\frac{8x^4}{(x^2+1)^3}-2\int_0^1\frac{t^2+1}{(x^2+1)^2}y(t)\,dt$$

with the mixed conditions $y(0)=1$ and $y(1)=\frac{1}{2}$. The exact solution of this problem is $(x^2+1)^{(-1)}$. We apply the suggested method with $J=1$ and $J=2$. The behavior of the approximate solution using the proposed method with $J=1$, $J=2$ and the exact solution are presented in Figure 1. InTable 1, we give the errors E_J of matrix A for different values of J. From this figure, it is clear that the proposed method can be considered as an efficient method to solve the linear integral equations. From Table 1 we see the errors decrease rapidly as J increase.

In Table 2 we compare the new method with $J=1$, $J=2$ and $J=3$ together with the exact solution. For the purpose of comparison we defined the meximum error for $u_J(x)$ as

$$E_J = \|u_J(x)-u(x)\|_\infty = \max\{|u_J(x)-u(x)|, a \le x \le b\}$$

Example 5 Consider the following second-order the Fredholm Linear Integro-Differential Equation

$$x^2u''(x)+50xu' = \frac{1-e^{(x+1)}}{x+1}\left(x^2+50x-35\right)e^x + \int_0^1 e^{xt}u(t)dt$$

with the initial conditions $u(0)=1$, $u(1)=e$ and exact solution e^x. This problem is solved by the same me- thods applied in example (4). Results are shown in Figure 2. From this figure, it is clear that the proposed method can be considered as an efficient method to solve the linear integral equations. For the purpose of com- parison in Table 3 we give the errors E_J of matrix A for different values of J. From Table 4 we see the errors decrease rapidly as J increase. In Table 4 we compare the new method with J = 1, J = 2 and J = 3 together with the exact solution.

Table 1. The maximum error matrix A from Example 4.

J	$J=1$	$J=2$	$J=3$	$J=4$	$J=5$
E_J	2.5×10^{-3}	3.5×10^{-5}	5.2×10^{-6}	2.6×10^{-7}	5.8×10^{-8}

Table 2. Error analysis and numerical results of Example 4.

t_i	Exact solution	u_j for $J=1$	u_j for $J=2$	u_j for $J=3$
0	$1.00000000e-00$	$1.00000000e-00$	$1.00000000e-00$	$1.00000000e-00$
0.125	$9.84643786e-01$	$9.81250761e-01$	$9.84057813e-01$	$9.84643237e-01$
0.2500	$9.41218794e-01$	$9.4075171e-01$	$9.41102894e-01$	$9.41218075e-01$
0.3750	$8.767335484e-01$	$8.76170942e-01$	$8.76731207e-01$	$8.76730782e-01$
0.5000	$8.00001246e-01$	$8.00180793e-01$	$8.00080476e-01$	$8.00001107e-01$
0.6250	$7.19101507e-01$	$7.18017064e-01$	$7.19852079e-01$	$7.19101672e-01$
0.7500	$6.40018961e-01$	$6.31094827e-01$	$6.4180973e-01$	$6.40018843e-01$
0.8750	$5.664334082e-01$	$5.50789164e-01$	$5.66027815e-01$	$5.66434729e-01$

Table 3. The maximum error matrix A from Example 5.

J	$J=1$	$J=2$	$J=3$	$J=4$	$J=5$
E_J	2.9×10^{-5}	5.2×10^{-6}	5.9×10^{-7}	2.2×10^{-7}	1.2×10^{-8}

Table 4. Error analysis and numerical results of Example 5.

t_i	Exact solution	u_j for $J=1$	u_j for $J=2$	u_j for $J=3$
0	1.00000000	1.00104921	1.00002430	1.00000013
0.125	1.13310019	1.13048310	1.13307891	1.13310007
0.2500	1.28400186	1.28178029	1.28314720	1.28400134
0.3750	1.45500374	1.45304829	1.45456781	1.45500352
0.5000	1.64870391	1.64910273	1.64820756	1.64870372
0.6250	1.86820582	1.86705912	1.86835405	1.86820565
0.7500	2.11704076	2.11078316	2.11527804	2.11704052
0.8750	2.39893108	2.39719802	2.39884240	2.39893214

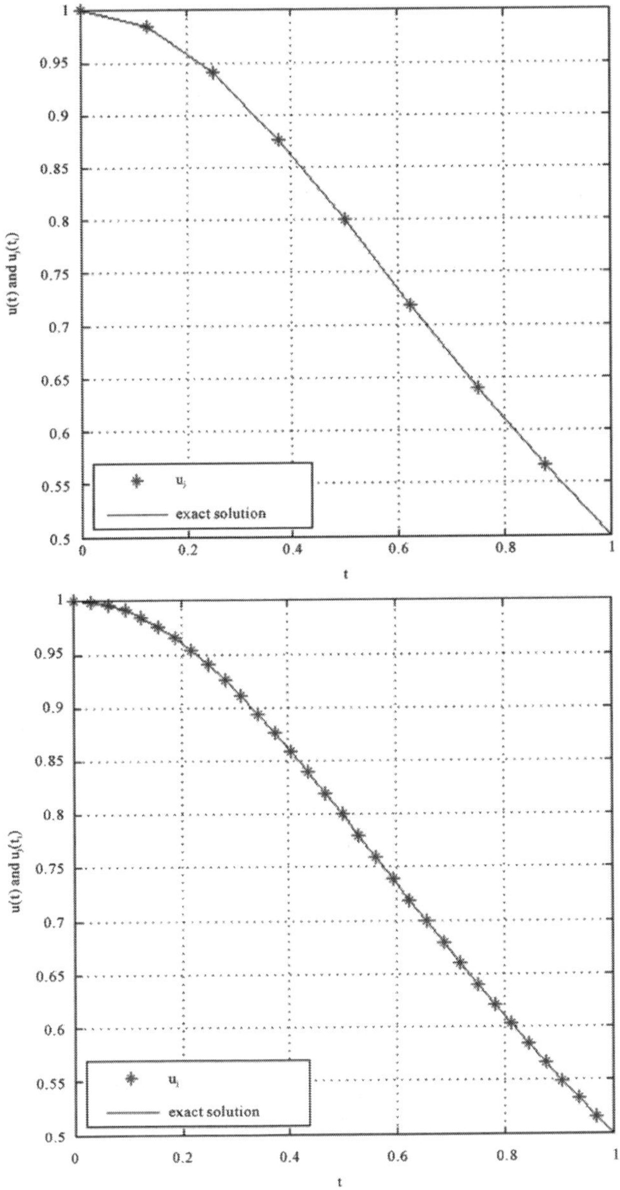

Figure 1. Result EX.4 for J = 1 and N = 7; Result EX.4 for J = 2 and N = 7.

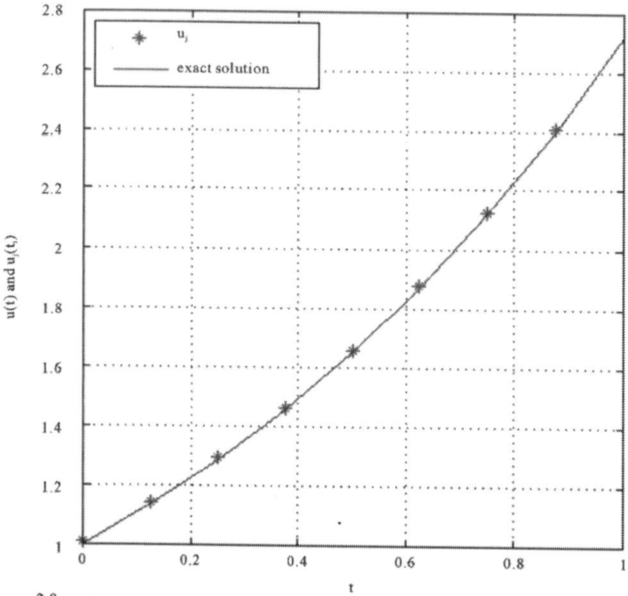

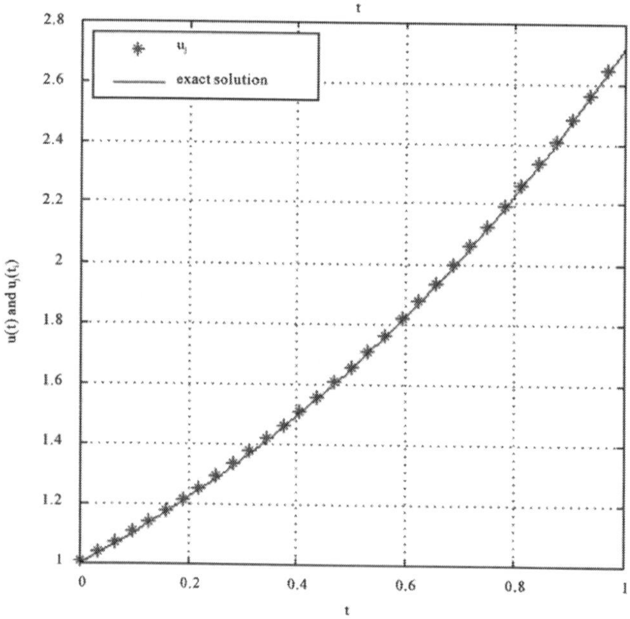

Figure 2. Result EX.5 for J = 1 and N = 7; Result EX.5 for J = 2 and N = 7.

5. CONCLUSION

Our results indicate that the method with the trigonometric scaling bases can be regarded as a structurally simple algorithm that is conventionally applicable to the numerical solution of IDEs. In addition, although we have re- stricted our attention to linear Fredholm IDEs, we expect the method to be easily extended to more general IDEs. the presented method which is based on the trigonometric scaling function is proposed to find the approximate solution. A comparison of the exact solution reveals that the presented method is very effective and convenient. Nevertheless, asFigure 1 and Figure 2 illustrate, the error of the trigonometric scaling bases shows that the accuracy improves with increasing J, hence for better results, using number J is recommended. Also form the obtained approximate solution, we can conclude that the proposed method gives the solution in an excellent agree- ment with the exact solution. All computations are done using MATLAB programming.

ACKNOWLEDGEMENTS

Hamid Safdari, Yones Esmaeelzade Aghdam The authors are very grateful to the editor for carefully reading the paper and for their comments and sugges- tions which have improved the paper.

REFERENCES

1. Desmond, R.A., Weiss, H.L., Arani, R.B., Soong, S.-J., Wood, M.J., Fiddian, P., Gnann, J. and Whitley, R.J. (2002) Clinical Applications for Change-Point Analysis of Herpes Zoster Pain. Journal of Pain and Symptom Management, 23, 510-516.

2. Wazwaz, D.D., Dimitrova, M.B. and Dishliev. A.B. (2000) Oscillation of the Bounded Solutions of Impulsive Differential-Difference Equations of Second Order. Applied Mathematics and Computation, 114, 61-68.

3. Gulsu, M. and Sezer, M. (2006) A Taylor Polynomial Approach for Solving Differential-Difference Equations. Journal of Computational and Applied Mathematics, 186, 349-369.

4. Maleknejad, K., Mirzaee, F. and Abbasbandy, S. (2004) Solving Linear Integro-Differential Equations System by Using Rationalized Haar Functions Method. Applied Mathematics and Computation, 155, 317-328.

5. Maleknejad, K., Tavassoli, M. and Kajani, M. (2004) Solving Linear Integro-Differential Equation by Galerkin Methods with Hybrid Functions. Journal of Computational and Applied Mathematics, 159, 603-612.

6. Ganesh, M. and Sloan, I.H. (1999) Optimal Order Spline Methods for Nonlinear Differential and Integro-Differential Equations. Applied Numerical Mathematics, 29, 445-478.

7. Dascoglua, A. and Sezer, M. (2005) Chebyshev Polynomial Solutions of Systems of Higher-Order Linear Fredholm-Volterra Integro-Differential Equations. Journal of the Franklin Institute, 342, 688-701.

8. Faour, A.L. and Saeed, R.K. (2006) Solution of a System of Linear Volterra Integral and Integro-Differential Equations by Spectral Method. Journal of Al-Nahrain University/Science, 62, 30-46.

9. Danfu, H. and Xufeng, A.S. (2007) Numerical Solution of Integro-Differential Equations by Appling CAS Wavelet Operational Matrix of Integration. Applied Mathematics and Computation, 194, 460-466.

10. Baker, C. and Tang, A. (1997) Stability of Continuous Implicit Runge-Kutta Methods for Volterra Integro-Differential Systems with Unbounded Delays. Applied Numerical Mathematics, 24, 153-173.

11. El-Sayed, S., Kaya, D. and Zarea, S. (2004) The Decomposition Method Applied to Solve High Order Linear Volterra-Fredholm Integro-Differential Equations. International Journal of Nonlinear Sciences and Numerical Simulation, 52, 105-112.

12. Sezer, M. and Gulsu, M. (2007) Polynomial Solution of the Most General Linear Fredholm-Volterra Integro Differential-Difference Equations by Means of Taylor Collocation Method. Applied Mathematics and Computation, 185, 646-657.

13. Sezer, M. and Gulsu, M. (2005) A New Polynomial Approach for Solving Difference and Fredholm Integro-Difference Equations with Mixed Argument. Applied Mathematics and Computation, 171, 332-344.

14. Chui, C.K. and Mhaskar, H.N. (1993) On Trigonometric Wavelets. Constructive Approximation, 9, 167-190.

15. Prestin, J. (2001) Trigonometric Wavelets. In: Jain, P.K., et al., Eds., Wavelet and Allied Topics, Narosa Publishing House, New Delhi, 183-217.

16. Themistoclakis, W. (1999) Trigonometric Wavelet Interpolation in Besov Spaces. Facta Univ. (Nis) Ser. Math. Inform, 14, 49-70.

17. Quak, E. (1996) Trigonometric Wavelets for Hermite Interpolation. Department of Mathematics, Texas A. M. University, 65, 683-722.

18. Chen, W.S. and Lin, W. (1997) Hadamard Singular Integral Equations and Its Hermite Wavelet Methods. Proceedings of the 5th International Colloquium on Finite Dimensional Complex Analysis, Beijing, 13-22.

19. Chen, W.S. and Lin, W. (2002) Trigonometric Hermite Wavelet and Natural Integral Equations for Stockes Problem. International Conference on Wavelet Analysis and Its Applications, Guangzhou, 73-86.

20. Lakestani, M. and Saray, B.N. (2010) Numerical Solution of Telegraph Equation Using Interpolating Scaling Functions. Computer and Mathematics with Application, 60, 1964-1972.

5

CHAPTER

Fractional Weierstrass Function by Application of Jumarie Fractional Trigonometric Functions and Its Analysis

Uttam Ghosh[1], Susmita Sarkar[2], Shantanu Das[3]

[1]Department of Mathematics, Nabadwip Vidyasagar College, Nabadwip, India

2Department of Applied Mathematics, University of Calcutta, Kolkata, India

3Reactor Control Systems Design Section, E & I Group, BARC, Mumbai, India

ABSTRACT

The classical example of no-where differentiable but everywhere continuous function is Weierstrass function. In this paper we have defined fractional order Weierstrass function in terms of Jumarie fractional trigonometric functions. The Hölder exponent and Box dimension of this new function have been evaluated here. It has been established that the values of Hölder exponent and Box dimension of this fractional order Weierstrass function are the same as in the original Weierstrass function. This new development in generalizing the classical Weierstrass function by use of fractional trigonometric function analysis and fractional derivative of fractional Weierstrass function by Jumarie fractional derivative, establishes that roughness indices are invariant to this generalization.

Keywords: Hölder Exponent, Fractional Weierstrass Function, Box Dimension, Jumarie Fractional Derivative, Jumarie Fractional Trigonometric Function

1. INTRODUCTION

The concepts of fractional geometry, fractional dimensions are important branches of science to study the irregularity of a function, graph or signals [1] - [3] . On the other hand fractional calculus is another developing mathematical tool to study the continuous but non-differentiable functions (signals) where the conventional calculus fails [4] - [11] . Many authors are trying to relate the fractional derivative and fractional dimension [1] [12] - [15] . The functions which are continuous but non-differentiable in integer order calculus can be characterized in terms of fractional calculus and especially through Holder exponent [10] [16] . To study the no-where differentiable functions authors in [12] - [16] used different types of fractional derivatives. Jumarie [17] defined the fractional trigonometric functions in terms of Mittag-Leffler function and established different useful fractional trigonometric formulas. The fractional order derivatives of those functions were established in-terms of the Jumarie [17] [18] modified fractional order derivatives. In this paper we have defined the fractional order Weierstrass functions in terms of the fractional order sine function. The Hölder exponent and box-dimension (fractional dimension) of graph of this function have been obtained here. The fractional order derivative of this function has also established here. This is a new development in generalizing the classical Weierstrass function by usage of fractional trigonometric functions including the study of its character. The paper is organized as: Section 2 deals with description of Jumarie fractional derivative, Mittag-Leffler function of one and two parameter types; fractional trigonometric function of one and two parameter types and derivation of Jumarie fractional derivatives of those functions. In this section we also have derived some useful relations of fractional trigonometric functions which shall be used for our further calculations—in characterizing fractional Weierstrass function. We have continued this section by introducing Lipschitz Hölder exponent (LHE)—its definition, its relation to Hurst exponent and fractional dimension and also definition of Hölder continuity. The classical Weierstrass function has also been defined here. These Lipschitz Hölder exponent, Hurst exponent, and fractional dimension are basic parameters to indicate roughness index of a function or a graph. In Section 3 we have described the fractional Weierstrass function by generalizing the classical Weierstrass function by use of fractional sine trigonometric function. Subsequently we apply derived identities of fractional trigonometric functions to evaluate the properties of this new fractional Weierstrass function. In Section 4 we have done derivation of properties of fractional derivatives of fractional Weierstrass function, and concluded the paper with conclusion and references.

2. JUMARIE FRACTIONAL ORDER DERIVATIVE AND MITTAG-LEFFLER FUNCTION

a) Fractional Order Derivative of Jumarie Type

Jumarie [17] defined the fractional order derivative by modifying the Left Riemann-Liouvellie (RL) fractional derivative in the following form for the function $f(x)$ in the interval a to x, with $f(x)=0$ for $x<a$.

$$
{}_{a}^{J}D_{x}^{\alpha}\left[f(x)\right]=\begin{cases}\dfrac{1}{\Gamma(-\alpha)}\displaystyle\int_{a}^{x}(x-\tau)^{-\alpha-1}f(\tau)\mathrm{d}\tau, & \alpha<0.\\ \dfrac{1}{\Gamma(1-\alpha)}\dfrac{\mathrm{d}}{\mathrm{d}x}\displaystyle\int_{a}^{x}(x-\tau)^{-\alpha}\left[f(\tau)-f(a)\right]\mathrm{d}\tau, & 0<\alpha<1\\ \left(f^{(\alpha-m)}(x)\right)^{(m)}, & m\le\alpha<m+1.\end{cases}
$$

(1)

In the above definition, the first expression is just the Riemann-Liouvelli fractional integration; the second line is Riemann-Liouvelli fractional derivative of order $0<\alpha<1$ of offset function that is $f(x)-f(a)$. For $\alpha>1$, we use the third line; that is first we differentiate the offset function with order $0<(\alpha-m)<1$, by the formula of second line, and then apply whole m order differentiation to it. Here we chose integer m, just less than the real number α; that is $m\le\alpha<m+1$. In this paper we use symbol ${}_{0}^{J}D_{x}^{\alpha}$ to denote Jumarie fractional derivative operator, as defined above. In case the start point value $f(a)$ is un-defined, there we take finite part of the offset function as $f(x)-f(a^{+})$; for calculations. Note in the above Jumarie definition ${}_{0}D_{x}^{\alpha}[C]=0$, where C is constant function, otherwise in RL sense, the fractional derivative of a constant function is

${}_0D_x^\alpha[C] = C\frac{x^{-\alpha}}{\Gamma(1-\alpha)}$, that is a decaying power-law function. Also we purposely state that $f(x)=0$ for $x<0$ in order to have initialization function in case of fractional differ-integration to be zero, else results are difficult [9] .

b) Mittag-Leffler Function and Its Jumarie Type Fractional Derivative: One and Two Parameter Type

1) One Parameter Mittag-Leffler Function

The Mittag-Leffler function [19] - [22] of one parameter is denoted by $E_\alpha(x)$ and defined by

$$E_\alpha(x) = \sum_{k=0}^{\infty}\frac{x^k}{\Gamma(1+k\alpha)} \tag{2}$$

This function plays a crucial role in classical calculus for $\alpha=1$, for $\alpha=1$ it becomes the exponential function, that is $E_1(x)=\exp(x)$

$$\exp(x) = \sum_{k=0}^{\infty}\frac{x^k}{k!} \tag{3}$$

We now consider the Mittag-Leffler function in the following form in infinite series representation for $f(x)=E_\alpha(x^\alpha)$ for $x\geq 0$ and $f(x)=0$ for $x<0$ as;

$$E_\alpha\left(x^\alpha\right)=1+\frac{x^\alpha}{\Gamma(1+\alpha)}+\frac{x^{2\alpha}}{\Gamma(1+2\alpha)}+\frac{x^{3\alpha}}{\Gamma(1+3\alpha)}+\cdots \tag{4}$$

Then taking Jumarie fractional derivative of order $0<\alpha<1$ term by term for the above series we obtain the following by using the formula ${}_0^J D_x^\alpha\left[x^\beta\right]=\frac{\Gamma(1+\beta)}{\Gamma(1+\beta-\alpha)}x^{\beta-\alpha}$ and ${}_0^J D_x^\alpha[1]=0$

$$\begin{aligned}{}_0^J D_x^\alpha\left[E_\alpha\left(ax^\alpha\right)\right]&={}_0^J D_x^\alpha\left[1+\frac{ax^\alpha}{\Gamma(1+\alpha)}+\frac{a^2x^{2\alpha}}{\Gamma(1+2\alpha)}+\frac{a^3x^{3\alpha}}{\Gamma(1+3\alpha)}+\cdots\right]\\&=0+a+\frac{a^2x^\alpha}{\Gamma(1+\alpha)}+\frac{a^3x^{2\alpha}}{\Gamma(1+2\alpha)}+\frac{a^4x^{3\alpha}}{\Gamma(1+3\alpha)}+\cdots\\ \therefore {}_0^J D_x^\alpha\left[E_\alpha\left(ax^\alpha\right)\right]&=aE_\alpha\left(ax^\alpha\right)\end{aligned} \tag{5}$$

Like the exponential function; $E_\alpha\left(x^\alpha\right)$ play important role in fractional calculus. The function $E_\alpha\left(x^\alpha\right)$ is a fundamental solution of the Jumarie type fractional differential equation ${}_0D_y^\alpha[y]=y$, where ${}_0D_x^\alpha$ is Jumarie derivative operator as described above.

Jumarie in [18] established $E_\alpha\left(i(x+y)^\alpha\right)=E_\alpha\left(ix^\alpha\right)\times E_\alpha\left(iy^\alpha\right)$. We reproduce the Proof of the above relation. Let us consider a function $f(x)$ which satisfies the condition

$$f\left(\lambda x^{\alpha}\right) f\left(\lambda y^{\alpha}\right)=f\left(\lambda(x+y)^{\alpha}\right).$$

Differentiating both side with respect to x and y of α -order respectively we get the following.

First consider y a constant, and we fractionally differentiate w.r.t. x by Jumarie derivative

$${}_0^J D_x^{\alpha}\left[f\left(\lambda x^{\alpha}\right)\right] \times {}_0^J D_x^{\alpha}\left[\lambda x^{\alpha}\right] \times f\left(\lambda y^{\alpha}\right)={}_0^J D_x^{\alpha}\left[f\left(\lambda(x+y)^{\alpha}\right)\right] \times {}_0^J D_x^{\alpha}\left[\lambda(x+y)^{\alpha}\right] \times {}_0^J D_x^{\alpha}[x+y]$$

$${}_0^J D_x^{\alpha}\left[f\left(\lambda x^{\alpha}\right)\right]=f^{\alpha}\left(\lambda x^{\alpha}\right) \qquad {}_0^J D_x^{\alpha}\left[f\left(\lambda(x+y)^{\alpha}\right)\right]=f^{\alpha}\left(\lambda(x+y)^{\alpha}\right)$$

$$f^{\alpha}\left(\lambda x^{\alpha}\right) \times\left[(\lambda) \frac{\Gamma(1+\alpha) x^{\alpha-\alpha}}{\Gamma(1+\alpha-\alpha)}\right] \times f\left(\lambda y^{\alpha}\right)=f^{\alpha}\left(\lambda(x+y)^{\alpha}\right) \times\left[(\lambda) \frac{\Gamma(1+\alpha)(x+y)^{\alpha-\alpha}}{\Gamma(1+\alpha-\alpha)}\right] \times {}_0^J D_x^{\alpha}[x+y]$$

Now we consider x as constant and do the following steps

$$f\left(\lambda x^{\alpha}\right) \times {}_0^J D_y^{\alpha}\left[f\left(\lambda y^{\alpha}\right)\right] \times {}_0^J D_y^{\alpha}\left[\lambda y^{\alpha}\right]={}_0^J D_y^{\alpha}\left[f\left(\lambda(x+y)^{\alpha}\right)\right] \times {}_0^J D_y^{\alpha}\left[\lambda(x+y)^{\alpha}\right] \times {}_0^J D_y^{\alpha}[x+y]$$

$${}_0^J D_y^{\alpha}\left[f\left(\lambda y^{\alpha}\right)\right]=f^{\alpha}\left(\lambda y^{\alpha}\right) \qquad {}_0^J D_y^{\alpha}\left[f\left(\lambda(x+y)^{\alpha}\right)\right]=f^{\alpha}\left(\lambda(x+y)^{\alpha}\right)$$

$$f\left(\lambda x^{\alpha}\right) \times f^{\alpha}\left(\lambda y^{\alpha}\right) \times\left[(\lambda) \frac{\Gamma(1+\alpha) y^{\alpha-\alpha}}{\Gamma(1+\alpha-\alpha)}\right]=f^{\alpha}\left(\lambda(x+y)^{\alpha}\right) \times\left[(\lambda) \frac{\Gamma(1+\alpha)(x+y)^{\alpha-\alpha}}{\Gamma(1+\alpha)}\right] \times {}_0^J D_y^{\alpha}[x+y]$$

Here we put equivalence of ${}_0^J D_y^{\alpha}[x+y] \equiv {}_0^J D_y^{\alpha}[x+y] \equiv {}_0^J D_u^{\alpha}[u+C]$, with C as constant; that is when x or y are taken as constant the function form of these two quantities gets equivalent that is equivalent to ${}_0^J D_u^{\alpha}[u]$ as

Jumarie fractional derivative of constant is zero. Therefore the RHS of above two expressions are equal, from that we get the following

$$f^{\alpha}\left(\lambda x^{\alpha}\right)f\left(\lambda y^{\alpha}\right)=f\left(\lambda x^{\alpha}\right)f^{\alpha}\left(\lambda y^{\alpha}\right)$$

$$\frac{f^{\alpha}\left(\lambda x^{\alpha}\right)}{f\left(\lambda x^{\alpha}\right)}=\frac{f^{\alpha}\left(\lambda y^{\alpha}\right)}{f\left(\lambda y^{\alpha}\right)}$$

The above two may be equated to a constant say λ. Then we have $f^{\alpha}\left(\lambda x^{\alpha}\right)=\lambda f\left(\lambda x^{\alpha}\right)$, or we write ${}_{0}^{J}D_{x}^{\alpha}\left[f\left(\lambda x^{\alpha}\right)\right]=\lambda f\left(\lambda x^{\alpha}\right)$. From the property of Mittag-Leffler function and Jumarie derivative of the Mittag- Leffler function we know that ${}_{0}^{J}D_{x}^{\alpha}\left[E_{\alpha}\left(ax^{\alpha}\right)\right]=aE_{\alpha}\left(ax^{\alpha}\right)$; we imply that the solution of $f^{\alpha}\left(\lambda x^{\alpha}\right)=\lambda f\left(\lambda x^{\alpha}\right)$ is $f\left(\lambda x^{\alpha}\right)=E_{\alpha}\left(\lambda x^{\alpha}\right)$.

Therefore $E_{\alpha}\left(\lambda x^{\alpha}\right)$ satisfies the condition $f\left(\lambda x^{\alpha}\right)f\left(\lambda y^{\alpha}\right)=f\left(\lambda(x+y)^{\alpha}\right)$, or $E_{\alpha}\left(\lambda x^{\alpha}\right)E_{\alpha}\left(\lambda y^{\alpha}\right)=E_{\alpha}\left(\lambda(x+y)^{\alpha}\right)$. Considering $\lambda=i$, we therefore can write the following identity

$$E_{\alpha}\left(i\left((x+y)^{\alpha}\right)\right)=E_{\alpha}\left(i\left(x^{\alpha}\right)\right)E_{\alpha}\left(i\left(y^{\alpha}\right)\right).$$

Using definition $E_{\alpha}\left(ix^{\alpha}\right)=\cos_{\alpha}\left(x^{\alpha}\right)+i\sin\left(x^{\alpha}\right)$ we expand the above as depicted below

$$\cos_\alpha (x+y)^\alpha + i\sin_\alpha (x+y)^\alpha = \left[\cos_\alpha (x^\alpha) + i\sin_\alpha (x^\alpha)\right] \times \left[\cos_\alpha (y^\alpha) + i\sin_\alpha (y^\alpha)\right]$$
$$= \left[\cos_\alpha (x^\alpha)\cos_\alpha (y^\alpha) - \sin_\alpha (y^\alpha)\sin_\alpha (x^\alpha)\right]$$
$$+ i\left[\sin_\alpha (x^\alpha)\cos_\alpha (y^\alpha) + \sin_\alpha (y^\alpha)\cos_\alpha (x^\alpha)\right].$$

Comparing real and imaginary part in above derived relation we get the following

$$\sin_\alpha (x+y)^\alpha = \sin_\alpha (x^\alpha)\cos_\alpha (y^\alpha) + \sin_\alpha (y^\alpha)\cos_\alpha (x^\alpha)$$
$$\cos_\alpha (x+y)^\alpha = \cos_\alpha (x^\alpha)\cos_\alpha (y^\alpha) - \sin_\alpha (y^\alpha)\sin_\alpha (x^\alpha)$$

This is very useful relation as in conjugation with classical trigonometric functions, and we will be using these relations in our analysis of fractional Weierstrass function and its fractional derivative.

2) Two Parameter Mittag-Leffler Function

The other important function is the two parameter Mittag-Leffler function denoted by $E_{\alpha,\beta}(x)$ and defined by,

$$E_{\alpha,\beta}(x) = \sum_{k=0}^{\infty} \frac{x^k}{\Gamma(\beta + k\alpha)} \qquad (6)$$

The functions (2) and (6) play important role in fractional calculus, also we note that $E_{\alpha,1}(x) = E_\alpha(x)$. Again from

Jumarie definition of fractional derivative we have ${}_0^J D_x^\alpha[1]=0$ and ${}_0^J D_x^\alpha\left[x^\beta\right]=\frac{\Gamma(1+\beta)}{\Gamma(1+\beta-\alpha)}x^{\beta-\alpha}$.

Again we derive Jumarie derivative of order β for one parameter Mittag-Leffler function $E_\alpha\left(x^\alpha\right)$ and thereby get two parameter Mittag-Leffler function. For finding term by term Jumarie derivative we use ${}_0^J D_x^\alpha\left[x^\upsilon\right]=\frac{\Gamma(1+\upsilon)}{\Gamma(1+\upsilon-\alpha)}x^{\upsilon-\alpha}$ and ${}_0^J D_x^\alpha[1]=0$.

$$
\begin{aligned}
{}_0^J D_x^\beta\left[E_\alpha\left(x^\alpha\right)\right] &= {}_0^J D_x^\beta\left[1+\frac{x^\alpha}{\Gamma(1+\alpha)}+\frac{x^{2\alpha}}{\Gamma(1+2\alpha)}+\frac{x^{3\alpha}}{\Gamma(1+3\alpha)}+\cdots\right] \\
&= 0+\frac{x^{\alpha-\beta}}{\Gamma(1+\alpha-\beta)}+\frac{x^{2\alpha-\beta}}{\Gamma(1+2\alpha-\beta)}+\frac{x^{3\alpha-\beta}}{\Gamma(1+3\alpha-\beta)}+\cdots=x^{\alpha-\beta}E_{\alpha,\alpha-\beta+1}\left(x^\alpha\right)
\end{aligned}
\tag{7}
$$

where $E_{\alpha,\alpha-\beta+1}\left(x^\alpha\right)$ is two parameter Mittag-Leffler function.

c) Jumarie Definition of Fractional Sine and Cosine Function and Their Fractional Derivative: Both One Parameter and Two Parameter Type

1) One Parameter Sine and Cosine Function

Jumarie [18] defined the one parameter fractional sine and cosine function in the following form,

$$E_\alpha\left(ix^\alpha\right)\overset{\text{def}}{=}\cos_\alpha\left(x^\alpha\right)+i\sin_\alpha\left(x^\alpha\right) \quad \text{(8a)}$$

$$\cos_\alpha\left(x^\alpha\right)\overset{\text{def}}{=}\sum_{k=0}^{\infty}(-1)^k\frac{x^{2k\alpha}}{\Gamma(1+2\alpha k)} \quad \text{(8b)}$$

$$\sin_\alpha\left(x^\alpha\right)\overset{\text{def}}{=}\sum_{k=0}^{\infty}(-1)^k\frac{x^{(2k+1)\alpha}}{\Gamma\left(1+(1+2k)\alpha\right)} \quad \text{(8c)}$$

From Figure 1 and Figure 2 it is observed that for $\alpha<1$ both the fractional trigonometric functions $\sin_\alpha\left(x^\alpha\right)$ and $\cos_\alpha\left(x^\alpha\right)$ is decaying functions like damped oscillatory motion. For $\alpha=1$ it is like simple harmonic motion with sustained oscillations; and for $\alpha>1$ it grows while it oscillates infinitely; like unstable oscillator.

The series representation of $f(t)=\cos_\alpha\left(t^\alpha\right)$ for $t\geq 0$ and $f(t)=0$ for $t<0$ is following

$$\cos_\alpha\left(at^\alpha\right)=1-\frac{a^2t^{2\alpha}}{\Gamma(1+2\alpha)}+\frac{a^4t^{4\alpha}}{\Gamma(1+4\alpha)}-\frac{a^6t^{6\alpha}}{\Gamma(1+6\alpha)}+\cdots$$

Taking term by term Jumarie derivative we get,

$$
\begin{aligned}
{}_0^J D_t^\alpha\left[\cos_\alpha\left(at^\alpha\right)\right] &= 0 - a^2\frac{\Gamma(1+2\alpha)t^{2\alpha-\alpha}}{\Gamma(1+2\alpha)\Gamma(1+\alpha)} + a^4\frac{\Gamma(1+4\alpha)t^{4\alpha-\alpha}}{\Gamma(1+4\alpha)\Gamma(1+3\alpha)} - a^6\frac{\Gamma(1+6\alpha)t^{6\alpha-\alpha}}{\Gamma(1+6\alpha)\Gamma(1+5\alpha)} + \cdots \\
&= -a\left[\frac{at^\alpha}{\Gamma(1+\alpha)} - \frac{a^3t^{3\alpha}}{\Gamma(1+3\alpha)} + \cdots\right]
\end{aligned}
$$

$$
\therefore {}_0^J D_t^\alpha\left[\cos_\alpha\left(at^\alpha\right)\right] = -a\sin_\alpha\left(at^\alpha\right)
\tag{9}
$$

The series presentation of $f(t) = \sin_\alpha\left(t^\alpha\right)$, for $t \geq 0$ with $f(t) = 0$ for $t < 0$ is

$$
\sin_\alpha\left(at^\alpha\right) = \frac{at^\alpha}{\Gamma(1+\alpha)} - \frac{a^3t^{3\alpha}}{\Gamma(1+3\alpha)} + \frac{a^5t^{5\alpha}}{\Gamma(1+5\alpha)} - \frac{a^7t^{7\alpha}}{\Gamma(1+7\alpha)} + \cdots
$$

Taking term by term Jumarie derivative we get

$$
\begin{aligned}
{}_0^J D_t^\alpha\left[\sin_\alpha\left(at^\alpha\right)\right] &= a\frac{\Gamma(1+\alpha)t^{\alpha-\alpha}}{\Gamma(1+\alpha)\Gamma(1+\alpha-\alpha)} - a^3\frac{\Gamma(1+3\alpha)t^{3\alpha-\alpha}}{\Gamma(1+3\alpha)\Gamma(1+3\alpha-\alpha)} \\
&\quad + a^5\frac{\Gamma(1+5\alpha)t^{5\alpha-\alpha}}{\Gamma(1+5\alpha)\Gamma(1+5\alpha-\alpha)} - a^7\frac{\Gamma(1+7\alpha)t^{7\alpha-\alpha}}{\Gamma(1+7\alpha)\Gamma(1+7\alpha-\alpha)} + \cdots \\
&= a\left[1 - \frac{a^2t^{2\alpha}}{\Gamma(1+2\alpha)} + \frac{a^4t^{4\alpha}}{\Gamma(1+4\alpha)} - \frac{a^6t^{6\alpha}}{\Gamma(1+6\alpha)} + \cdots\right]
\end{aligned}
$$

$$
\therefore {}_0^J D_t^\alpha\left[\sin_\alpha\left(at^\alpha\right)\right] = a\cos_\alpha\left(t^\alpha\right)
\tag{10}
$$

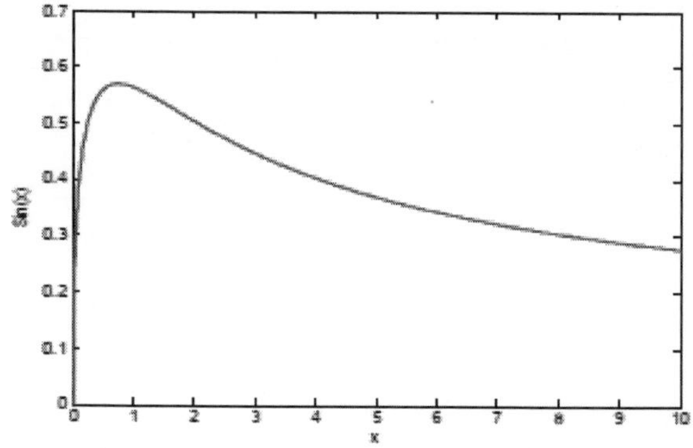
0.7
0.6
0.5
0.4
0.3
0.2
0.1
0
Sin(x)
0 1 2 3 4 5 6 7 8 9 10
x

(a)

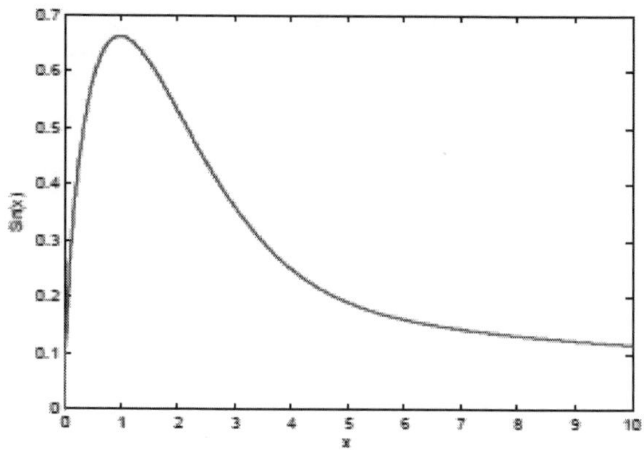
0.7
0.6
0.5
0.4
0.3
0.2
0.1
0
Sin(x)
0 1 2 3 4 5 6 7 8 9 10
x

(b)

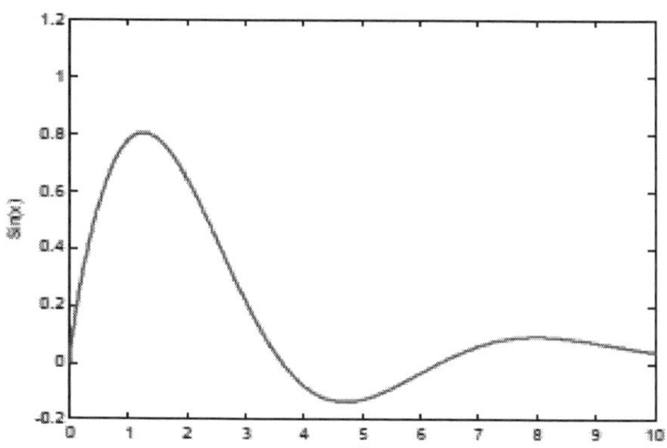
1.2
1
0.8
0.6
0.4
0.2
0
-0.2
Sin(x)
0 1 2 3 4 5 6 7 8 9 10
x

(c)

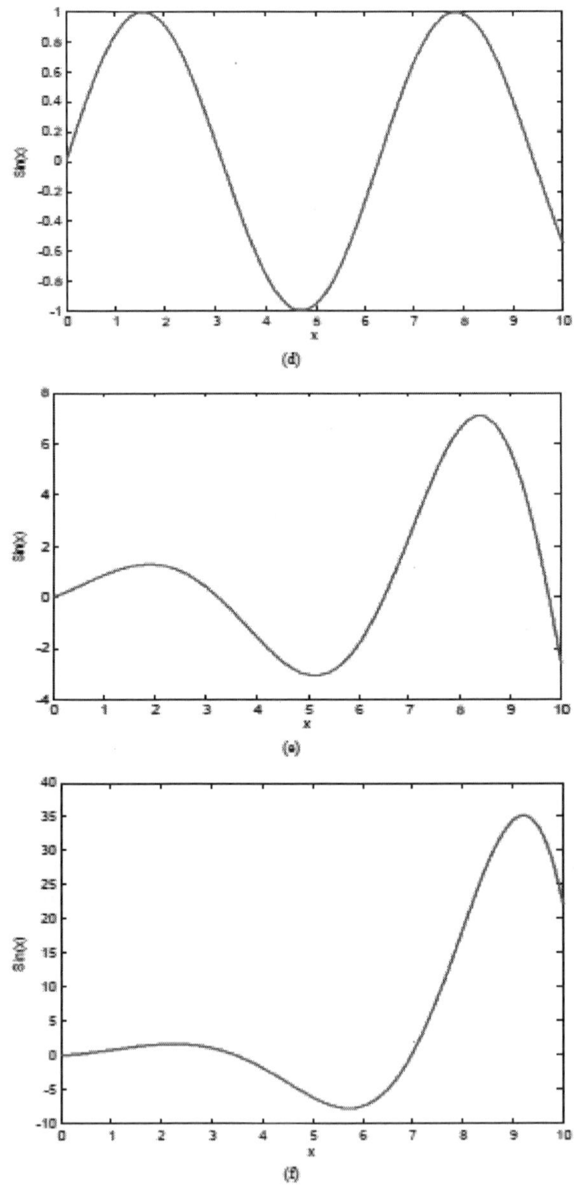

Figure 1. Graph of $\sin_\alpha(x^\alpha)$. (a) For $\alpha = 0.4$; (b) For $\alpha = 0.6$; (c) For $\alpha = 0.8$; (d) For $\alpha = 1.0$; (f) For $\alpha = 1.2$; (g) For $\alpha = 1.4$.

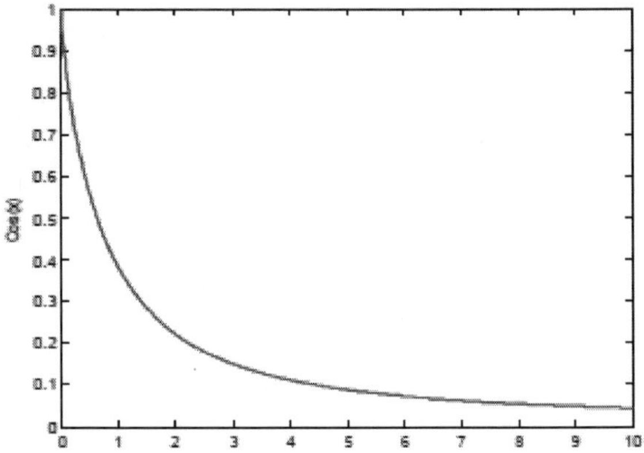
0
0.1
0.2
0.3
0.4
0.5
0.6
0.7
0.8
0.9
1
0 1 2 3 4 5 6 7 8 9 10
x

(a)

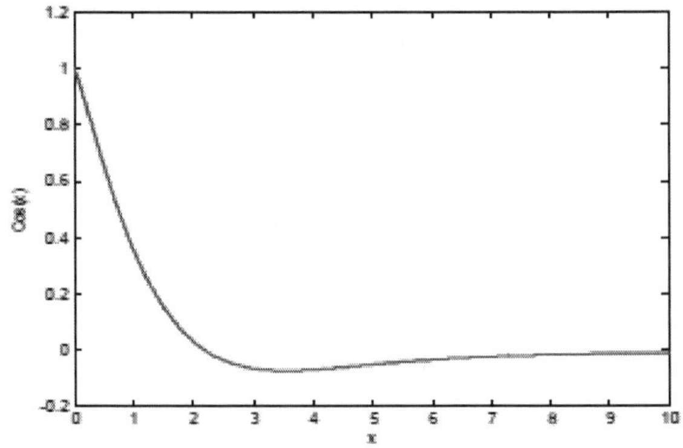
-0.2
0
0.2
0.4
0.6
0.8
1
1.2
0 1 2 3 4 5 6 7 8 9 10
x

(b)

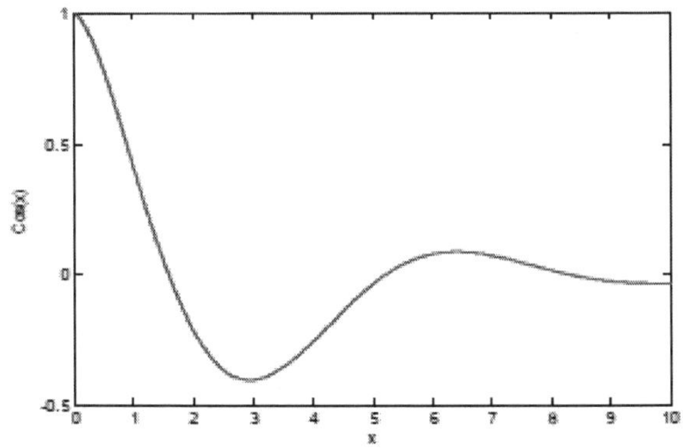
-0.5
0
0.5
1
0 1 2 3 4 5 6 7 8 9 10
x

(c)

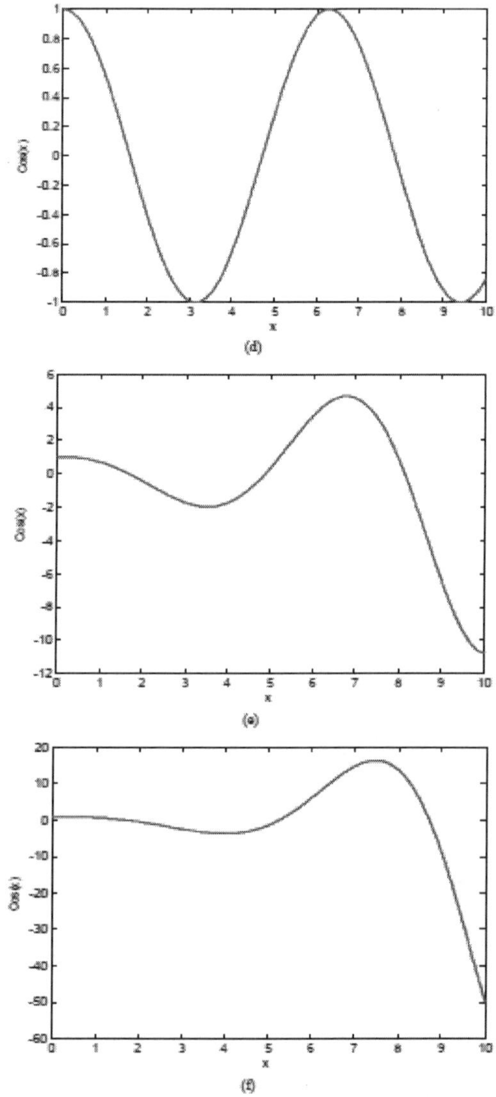

Figure 2. Graph of $\cos_\alpha\left(x^\alpha\right)$. (a) For $\alpha = 0.4$; (b) For $\alpha = 0.6$; (c) For $\alpha = 0.8$; (d) For $\alpha = 1.0$; (e) For $\alpha = 1.2$; (f) For $\alpha = 1.4$.

Thus we get

$${}_0^J D_x^\alpha\left[\cos_\alpha\left(ax^\alpha\right)\right]=-a\sin_\alpha\left(x^\alpha\right) \text{ and } {}_0^J D_x^\alpha\left[\sin_\alpha\left(ax^\alpha\right)\right]=a\cos_\alpha\left(x^\alpha\right).$$

2) Two Parameter Sine and Cosine Function

Let us define the two parameter sine and cosine functions $\cos_{\alpha,\beta}\left(x^\alpha\right)$ and $\sin_{\alpha,\beta}\left(x^\alpha\right)$ as depicted below:

$$\cos_{\alpha,\beta}\left(x^\alpha\right)\overset{\text{def}}{=}\sum_{k=0}^{\infty}(-1)^k\frac{x^{2k\alpha}}{\Gamma(\beta+2\alpha k)}=\frac{1}{\Gamma(\beta)}-\frac{x^{2\alpha}}{\Gamma(2\alpha+\beta)}+\frac{x^{4\alpha}}{\Gamma(4\alpha+\beta)}-\cdots$$

$$\sin_{\alpha,\beta}\left(x^\alpha\right)\overset{\text{def}}{=}\sum_{k=0}^{\infty}(-1)^k\frac{x^{(2k+1)\alpha}}{\Gamma(\beta+(1+2k)\alpha)}=\frac{x^{\alpha}}{\Gamma(\alpha+\beta)}-\frac{x^{3\alpha}}{\Gamma(3\alpha+\beta)}+\frac{x^{5\alpha}}{\Gamma(5\alpha+\beta)}+\cdots$$

Now with this and with definition of two parameter Mittag-Leffler function (3) with imaginary argument we get the following useful identity

$$\begin{aligned}E_{\alpha,\beta}\left(ix^\alpha\right)&=\sum_{k=0}^{\infty}\frac{\left(ix^\alpha\right)^k}{\Gamma(\beta+k\alpha)}=\frac{1}{\Gamma(\beta)}+\frac{(ix)^\alpha}{\Gamma(\alpha+\beta)}+\frac{\left(ix^\alpha\right)^2}{\Gamma(2\alpha+\beta)}+\frac{\left(ix^\alpha\right)^3}{\Gamma(3\alpha+\beta)}+\cdots\\&=\left(\frac{1}{\Gamma(\beta)}-\frac{x^{2\alpha}}{\Gamma(2\alpha+\beta)}+\frac{x^{4\alpha}}{\Gamma(4\alpha+\beta)}-\cdots\right)+i\left(\frac{x^{\alpha}}{\Gamma(\alpha+\beta)}-\frac{x^{3\alpha}}{\Gamma(3\alpha+\beta)}+\frac{x^{5\alpha}}{\Gamma(5\alpha+\beta)}+\cdots\right)\\&=\cos_{\alpha,\beta}\left(x^\alpha\right)+i\sin_{\alpha,\beta}\left(x^\alpha\right)\end{aligned}$$

Now for $\beta > \alpha$, we do the Jumarie derivative of order α on the function $f(x) = x^{\beta-1}\cos_{\alpha,\beta}(x^\alpha)$ as depicted in following steps, with formula ${}_0^JD_x^\alpha\left[x^\upsilon\right] = \dfrac{\Gamma(1+\upsilon)}{\Gamma(1+\upsilon-\alpha)}x^{\upsilon-\alpha}$ and ${}_0^JD_x^\alpha[1] = 0$.

$$\begin{aligned}
{}_0^JD_x^\alpha\left[x^{\beta-1}\cos_{\alpha,\beta}(x^\alpha)\right] &= {}_0^JD_x^\alpha\left[\left(x^{\beta-1}\right)\times\left(\frac{1}{\Gamma(\beta)} - \frac{x^{2\alpha}}{\Gamma(2\alpha+\beta)} + \frac{x^{4\alpha}}{\Gamma(4\alpha+\beta)} - \dots\right)\right] \\
&= \frac{x^{\beta-\alpha-1}}{\Gamma(\beta-\alpha)} - \frac{x^{\beta+\alpha-1}}{\Gamma(\beta+\alpha)} + \frac{x^{\beta+3\alpha-1}}{\Gamma(\beta+3\alpha)} - \dots \\
&= \left(x^{\beta-\alpha-1}\right)\times\left[\frac{1}{\Gamma(\beta-\alpha)} - \frac{x^{2\alpha}}{\Gamma(\beta-\alpha+2\alpha)} + \frac{x^{\beta+3\alpha-1}}{\Gamma(\beta-\alpha+4\alpha)} - \dots\right] \\
&= x^{\beta-\alpha-1}\cos_{\alpha,\beta-\alpha}(x^\alpha)
\end{aligned}$$

Thus we get a very useful relation

$${}_0^JD_x^\alpha\left[x^{\beta-1}\cos_{\alpha,\beta}(x^\alpha)\right] = x^{\beta-\alpha-1}\cos_{\alpha,\beta-\alpha}(x^\alpha).$$

Similarly it can be shown that

$${}_0^JD_x^\alpha\left[x^{\beta-1}\sin_{\alpha,\beta}(x^\alpha)\right] = x^{\beta-\alpha-1}\sin_{\alpha,\beta-\alpha}(x^\alpha).$$

Now we calculate the Jumarie type fractional order derivative of $\exp(x) = E_1(x)$ like we did for $E_\alpha(x^\alpha)$ by using the formula ${}_0^JD_x^\alpha\left[x^\upsilon\right] = \dfrac{\Gamma(1+\upsilon)}{\Gamma(1+\upsilon-\alpha)}x^{\upsilon-\alpha}$ and ${}_0^JD_x^\alpha[1] = 0$.

$$
{}_0^J D_x^\alpha \left[\exp(ax)\right] = {}_0^J D_x^\alpha \left[E_1(ax)\right] = {}_0^J D_x^\alpha \left(1 + \frac{ax}{\Gamma(2)} + \frac{a^2x^2}{\Gamma(3)} + \frac{a^3x^3}{\Gamma(4)} + \cdots\right)
$$

$$
= 0 + \frac{ax^{1-\alpha}}{\Gamma(2-\alpha)} + \frac{a^2x^{2-\alpha}}{\Gamma(3-\alpha)} + \frac{a^3x^{3-\alpha}}{\Gamma(4-\alpha)} + \cdots = aE_{1,2-\alpha}(ax)
$$

On the other hand the Jumarie type fractional order derivative of $\cos(ax)$ is following, as we did for $\cos_\alpha(x^\alpha)$ by using the formula ${}_0^J D_x^\alpha \left[x^\upsilon\right] = \frac{\Gamma(1+\upsilon)}{\Gamma(1+\upsilon-\alpha)} x^{\upsilon-\alpha}$ and ${}_0^J D_x^\alpha [1] = 0$.

$$
{}_0^J D_x^\alpha \left[\cos(ax)\right] = {}_0^J D_x^\alpha \left[1 - \frac{a^2x^2}{\Gamma(3)} + \frac{a^4x^4}{\Gamma(5)} + \frac{a^6x^6}{\Gamma(7)} + \cdots\right]
$$

$$
= 0 - \frac{a^2x^{2-\alpha}}{\Gamma(3-\alpha)} + \frac{a^4x^{4-\alpha}}{\Gamma(5-\alpha)} - \frac{a^6x^{6-\alpha}}{\Gamma(7-\alpha)} + \cdots
$$

$$
= -ax^{1-\alpha} \sin_{1,2-\alpha}(ax)
$$

We obtain

$$
{}_0^J D_x^\alpha \left[\cos(ax)\right] = -ax^{1-\alpha} \sin_{1,2-\alpha}(ax).
$$

Similarly the Jumarie type fractional order derivative of $\sin(x)$ is

$$
{}_0^J D_x^\alpha \left[\sin(ax)\right] = ax^{1-\alpha} \cos_{1,2-\alpha}(ax).
$$

2.1. Definition of Some Useful Roughness Indices

a) Lipschitz Hölder Exponent (LHE)

A function is said to have LHE [1] α it satisfies the following condition

$$|f(x)-f(y)| \sim |x-y|^{\alpha} \qquad 0<|x-y|<\varepsilon$$

where ε is a small positive number. The property LHE defined above corresponds to local property. The global LHE in interval $[a,b]$ is denoted by λ and is defined by

$$\lambda = \inf_{x\in[a,b]} \alpha$$

unless $f(x)$ is a constant function, $\lambda \le 1$. The Lipschitz Holder exponent is sometimes named as Holder exponent. For the continuous function $f: R \to R$, $f(x)$ satisfies the Lipschitz condition on its domain of definition if $|f(x)-f(y)|<C|x-y|$ when $0<|x-y|<\varepsilon$, where ε is small positive number, and $C>0$ is real constant. This function $f(x)$ has Holder exponent as unity.

Consider the function:

$f:\mathbb{R}\to\mathbb{R}$ such that $f(x)=\sin(x)$ then $|f(x)-f(y)|=|\sin(x)-\sin(y)|<C|x-y|$ when $0<|x-y|<\varepsilon$ is a function with Holder exponent 1. In a way it states that the continuous function in consideration is one-whole differentiable and the value of differentiation is bounded, that is $\frac{|f(x)-f(y)|}{|x-y|}<C$ for $0<|x-y|<\varepsilon$.

b) Holder Continuity

A continuous function $f(x)$ which is non-differentiable in classical sense is said to holder continuous with exponent α if

$$|f(x)-f(y)|<C|x-y|^{\alpha} \qquad 0<|x-y|<\varepsilon$$

where $C>0$ is a real constant and $\varepsilon>0$.

c) Fractional Dimension

Fractional dimension (d) or box dimension [1] of a function or graph is local property, denotes the degree of roughness of a function or graph. Let the graph of a function is $f(x)$ for $x\in[a,b]$ can be covered by N-squares of size r then with $\lim(r\to 0)$ the fractional dimension of the graph is defined as,

$$d=\lim_{r\to 0}\frac{\log(N)}{\log(1/r)}$$

Again if H be the Hurst exponent then the relation between the above Holder exponents are $d=2-H=2-\alpha$ [1] [9]. The Holder and Hurst exponents are equivalent for uni-fractal graphs that has a constant fractional dimension in defined interval [1] [9].

3. The Fractional Weierstrass Function

In 1872 K. Weierstrass [23] - [25] proposed his famous example of an everywhere continuous but no-where differentiable function $W(x)$ on the real line $\mathbb{R}$ with two parameters $b \geq a > 1$ in the following form

$$W(x) = \sum_{k=1}^{\infty} a^{-k} \sin\left(b^k x\right) \qquad x \in \mathbb{R}$$

where b is odd-integer. He proved that this function is continuous for all $x \in R$ and is non-differentiable for all real values of x provided $ab > 1 + \frac{3\pi}{2}$. Considering b a constant say $b = \lambda$ a constant and assuming, and $s = 2 - \frac{\log a}{\log b}$ another presentation of the Weierstrass function [13] can be obtained which is

$$W(x) = \sum_{k=1}^{\infty} \lambda^{(s-2)k} \sin\left(\lambda^k x\right) \qquad \lambda > 1 \quad 1 < s < 2 \tag{11}$$

In reference [13] Falconer established the fractional dimension of Weierstrass function defined in (11) is s and the corresponding Holder exponent is $2 - s$.

We define the fractional Weierstrass Function in terms of Jumarie [2008] fractional sine function, that is $\sin_\alpha\left(x^\alpha\right)$ in the following form for $x \geq 0$

$$W_\alpha\left(x^\alpha\right)=\sum_{k=1}^{\infty}\lambda^{(s-2)k}\sin_\alpha\left(\lambda^{\alpha k}x^\alpha\right) \qquad \lambda>1 \qquad 1<s<2$$

(12)

where, $0<\alpha<1$, and for $\alpha=1$ it reduces the original Weierstrass Function, and a condition that $W_\alpha\left(x^\alpha\right)=0$ for $x<0$.

We only are stating some lemmas which will be used to characterize the fractional Weierstrass function and its fractional derivative.

Lemma 1:

Let f be function continuous in interval $[0,1]$ and $0\le s\le 1$ [12] -[14] .

Suppose

1) $|f(x)-f(y)|\le C|x-y|^s \qquad 0<x \quad y<1$

then the dimension [12] - [14] of the graph f is $d\le 2-s$.

2) Suppose $\delta_0>0$. For every $x\in[0,1]$, and $0<\delta<\delta_0$ there exists $y\in[0,1]$ such that $|x-y|<\delta$ and $|f(x)-f(y)|\ge C\delta^s$ then the dimension [12] -[14] of the graph f is $d\ge 2-s$.

Theorem 1: The Holder exponent of fractional Weierstrass function $W_\alpha\left(x^\alpha\right)$ with $0<\alpha<1$ is $2-s$ and consequently the

Hausdorff dimension or fractional dimension is s over any finite interval suppose it is $[0,1]$.

Proof: We calculate $W_\alpha\left[(x+h)^\alpha\right]-W_\alpha\left[x^\alpha\right]$ in following steps where we have used our derived expression

$$\sin_\alpha\left(a(x+y)^\alpha\right)=\left[\sin_\alpha\left(ax^\alpha\right)\cos_\alpha\left(ay^\alpha\right)+\cos_\alpha\left(ax^\alpha\right)\sin_\alpha\left(ay^\alpha\right)\right]$$

$$W_\alpha\left[(x+h)^\alpha\right]-W_\alpha\left[x^\alpha\right]=\sum_{k=1}^{\infty}\lambda^{(s-2)k}\sin_\alpha\left(\lambda^{\alpha k}(x+h)^\alpha\right)-\sum_{k=1}^{\infty}\lambda^{(s-2)k}\sin_\alpha\left(\lambda^{\alpha k}(x)^\alpha\right)$$

$$=\sum_{k=1}^{\infty}\lambda^{(s-2)k}\left[\sin_\alpha\left(\lambda^{\alpha k}x^\alpha\right)\cos_\alpha\left(\lambda^{\alpha k}h^\alpha\right)+\cos_\alpha\left(\lambda^{\alpha k}x^\alpha\right)\sin_\alpha\left(\lambda^{\alpha k}h^\alpha\right)\right]-\sum_{k=1}^{\infty}\lambda^{(s-2)k}\sin_\alpha\left(\lambda^{\alpha k}(x)^\alpha\right)$$

$$=\sum_{k=1}^{\infty}\lambda^{(s-2)k}\left[\sin_\alpha\left(\lambda^{\alpha k}x^\alpha\right)\left(\cos_\alpha\left(\lambda^{\alpha k}h^\alpha\right)-1\right)+\cos_\alpha\left(\lambda^{\alpha k}x^\alpha\right)\sin_\alpha\left(\lambda^{\alpha k}h^\alpha\right)\right]$$

From the series expansion of $\sin_\alpha\left(\lambda^{\alpha k}x^\alpha\right)$ and $\cos_\alpha\left(\lambda^{\alpha k}x^\alpha\right)$ and also from the Figure 1 andFigure 2, it is clear that for small x, $\sin_\alpha\left(\lambda^{\alpha k}x^\alpha\right)\approx\lambda^{\alpha k}x^\alpha$ and $\cos_\alpha\left(\lambda^{\alpha k}x^\alpha\right)\approx 1$ also both $\left|\sin_\alpha\left(\lambda^{\alpha k}x^\alpha\right)\right|$ and $\left|\cos_\alpha\left(\lambda^{\alpha k}x^\alpha\right)\right|$ is less than or equal to 1. Therefore, with above observation that is for small h, $\sin_\alpha\left(\lambda^{\alpha k}h^\alpha\right)\approx\lambda^{\alpha k}h^\alpha$, $\cos_\alpha\left(\lambda^{\alpha k}h^\alpha\right)-1\approx 0$ and for large h, $\cos_\alpha\left(\lambda^{\alpha k}h^\alpha\right)\approx 0$ we write the following

$$\left|W_\alpha\left[(x+h)^\alpha\right]-W_\alpha\left[x^\alpha\right]\right|$$
$$\le\sum_{k=1}^{\infty}\lambda^{(s-2)k}\left[\left|\sin_\alpha\left(\lambda^{\alpha k}x^\alpha\right)\right|\left|\cos_\alpha\left(\lambda^{\alpha k}h^\alpha\right)-1\right|+\left|\cos_\alpha\left(\lambda^{\alpha k}x^\alpha\right)\right|\left|\sin_\alpha\left(\lambda^{\alpha k}h^\alpha\right)\right|\right]$$
$$\le\sum_{k=1}^{\infty}\lambda^{(s-2)k}\left[\min\left(\lambda^{\alpha k}h^\alpha,1\right)\right]$$

Choose $0<h<1$ then one can find positive integer m such that $\lambda^{-(m+1)}\le h\le\lambda^{-m}$ then divide the summation that is $\sum_{k=1}^{\infty}\lambda^{(s-2)k}\left[\min\left(\lambda^{\alpha k}h^\alpha,1\right)\right]$ into two parts. First part for $k=1$ to m then $\sin_\alpha\left(\lambda^{\alpha k}x^\alpha\right)\approx\lambda^{\alpha k}x^\alpha$ and for other values of k maximum value of the expression in third bracket is equal to 1. We use the geometric series formulas $\sum_{k=1}^{m}a^k=a\left(\frac{a^m-1}{a-1}\right)$ and $\sum_{k=1}^{\infty}a^k=\frac{a}{1-a}$,for $\sum_{k=m+1}^{\infty}x^k=\frac{x^{m+1}}{1-x}$ in the following derivation.

$$\left|W_\alpha\left[(x+h)^\alpha\right]-W_\alpha\left[x^\alpha\right]\right|\le\sum_{k=1}^{m}\lambda^{(s-2)k}\left(\lambda^{\alpha k}h^\alpha\right)+1\sum_{k=m+1}^{\infty}\lambda^{(s-2)k}$$
$$=h^\alpha\sum_{k=1}^{m}\lambda^{(s-2+\alpha)k}+\sum_{k=m+1}^{\infty}\lambda^{(s-2)k}$$
$$=h^\alpha\left(\lambda^{(s-2+\alpha)}\frac{\lambda^{(s-2+\alpha)m}-1}{\lambda^{(s-2+\alpha)}-1}\right)+\left(\frac{\lambda^{(s-2)(m+1)}}{1-\lambda^{(s-2)}}\right)$$
$$\le h^\alpha\frac{\lambda^{(s-2+\alpha)(m+1)}}{\lambda^{(s-2+\alpha)}-1}+\frac{\lambda^{(s-2)(m+1)}}{1-\lambda^{(s-2)}}$$

With $\lambda^{-(m+1)} \le h \le \lambda^{-m}$, that is $\lambda^{(m+1)} \ge h^{-1} \ge \lambda^{m}$ we get the following

$$\left|W_\alpha\left[(x+h)^\alpha\right]-W_\alpha\left[x^\alpha\right]\right| \le h^\alpha \frac{h^{-(s-2+\alpha)}}{\lambda^{(s-2+\alpha)}-1}+\frac{h^{-(s-2)}}{1-\lambda^{(s-2)}}$$
$$=\left(\frac{1}{\lambda^{(s-2+\alpha)}-1}+\frac{1}{1-\lambda^{(s-2)}}\right)h^{2-s}=C_1h^{2-s}$$

where the constant $C_1=\frac{1}{\lambda^{(s-2+\alpha)}-1}+\frac{1}{1-\lambda^{(s-2)}}$. From definition of Holderian function and the above discus-

sion it is clear that fractional Weierstrass function is also Holder continuous with Holder exponent $(2-s)$, a fractional number. This shows (by Lemma-1) that Hausdorff dimension of graph of fractional Weierstrass function is $[2-(2-s)]=s$. Thus the Hausdorff dimension of fractional Weierstrass function and original Weierstrass function is same, is independent of fractional exponent (α) as defined in (11).

4. THE JUMARIE FRACTIONAL DERIVATIVE OF FRACTIONAL WEIERSTRASS FUNCTION

Many authors found the fractional derivative of the continuous but nowhere differentiable function that is Weierstrass Function [10] -[17] using different type definitions of fractional derivatives. Here we consider Jumarie type fractional order derivative of $W_\alpha\left(x^\alpha\right)$ is of order α

$${}_0^J D_x^\alpha\left[W_\alpha\left(x^\alpha\right)\right]=\sum_{k=1}^{\infty}\lambda^{(s-2)k}\left({}_0^J D_x^\alpha\left[\sin_\alpha\left(\lambda^{\alpha k}x^\alpha\right)\right]\right)=\sum_{k=1}^{\infty}\lambda^{(s-2)k}\lambda^{\alpha k}\cos_\alpha\left(\lambda^{\alpha k}x^\alpha\right).$$

We used in above derivation the identity ${}_0^J D_x^\alpha\left[\sin_\alpha\left(ax^\alpha\right)\right]=a\cos_\alpha\left(ax^\alpha\right)$. Therefore from above derivation we obtain the following,

$${}_0^J D_x^\alpha\left[W_\alpha\left(x^\alpha\right)\right]=\sum_{k=1}^{\infty}\lambda^{(s-2+\alpha)k}\cos_\alpha\left(\lambda^{\alpha k}x^\alpha\right). \tag{5}$$

Since if $0<\alpha<1$ then $\cos_\alpha\left(\lambda^{\alpha k}x^\alpha\right)$ is a bounded function and therefore ${}_0^J D_x^\alpha\left[W_\alpha\left(x^\alpha\right)\right]$ will be bounded function if $\sum_{k=1}^{\infty}\lambda^{(s-2+\alpha)k}$ is convergent. Since $\sum_{k=1}^{\infty}\lambda^{(s-2+\alpha)k}$ is a geometric series will be conver-

gent if $s-2+\alpha<0$ implying $\alpha<2-s$. Hence the fractional derivative of order α with $0<\alpha<1$ of the Weierstrass Function will exists when $\alpha<2-s$.

Again if $\alpha>1$ then $\cos_\alpha\left(\lambda^{\alpha k}x^\alpha\right)$ and $\sin_\alpha\left(\lambda^{\alpha k}x^\alpha\right)$ for $k=1,2,3,\cdots$ are unbounded functions (Figure 1 and Figure 2) and will grow by oscillating without bound to $\pm\infty$ for $x\to\infty$. Since $1<s<2$ and $\alpha>1$ implying $s+\alpha-2>0$ therefore $\sum_{k=1}^{\infty}\lambda^{(s-2+\alpha)k}$ is a divergent series. Therefore

$$ {}_0^J D_x^\alpha \left[W_\alpha \left(x^\alpha \right) \right] = \sum_{k=1}^{\infty} \lambda^{(s-2+\alpha)k} \cos_\alpha \left(\lambda^{\alpha k} x^\alpha \right) $$

is a divergent series for $\alpha > 1$. We write following observation

$$ {}_0^J D_x^\alpha \left[W_\alpha \left(x^\alpha \right) \right] = \begin{cases} \text{Bounded for} & \alpha < 2-s \\ \text{Unbounded for} & \alpha \geq 2-s \end{cases} $$

This shows that α -order $(0 < \alpha < 1)$ Jumarie fractional derivative of the fractional Weierstrass function exists when $\alpha < 2-s$ and for $\alpha \geq 2-s$ it does not exist. Thus we can state a theorem in the following form

Theorem 2: α -order $(0 < \alpha < 1)$ Jumarie fractional derivative of the fractional Weierstrass function

$$ W_\alpha \left(x^\alpha \right) = \sum_{k=1}^{\infty} \lambda^{(s-2)k} \sin_\alpha \left(\lambda^{\alpha k} x^\alpha \right) \qquad \lambda > 1 \qquad 1 < s < 2 $$

exists when $\alpha < 2-s$ and for $\alpha \geq 2-s$ it does not exist.

Theorem 3: The Holder exponent of α -order fractional derivative of fractional Weierstrass function $W_\alpha \left(x^\alpha \right)$, $0 < \alpha < 1$ is $2-s-\alpha$ and consequently the Hausdorff dimension or fractional dimension is $s+\alpha$ over any finite interval $[0,1]$.

Proof: Let

$$ {}_0^J D_x^\alpha \left[W_\alpha \left(x^\alpha \right) \right] = W_\alpha^{(\alpha)} \left(x^\alpha \right) = \sum_{k=1}^{\infty} \lambda^{(s-2+\alpha)k} \cos_\alpha \left(\lambda^{\alpha k} x^\alpha \right) $$

denotes α-order fractional Jumarie derivative of fractional Weierstrass function. Then using the identity $\cos_\alpha \left(a(x+y)^\alpha \right) = \left[\cos_\alpha \left(ax^\alpha \right) \cos_\alpha \left(ay^\alpha \right) - \sin_\alpha \left(ax^\alpha \right) \sin_\alpha \left(ay^\alpha \right) \right]$ we get the following

$$ W_\alpha^{(\alpha)} \left[(x+h)^\alpha \right] - W_\alpha^{(\alpha)} \left[x^\alpha \right] = \sum_{k=1}^{\infty} \lambda^{(s-2+\alpha)k} \cos_\alpha \left(\lambda^{\alpha k} (x+h)^\alpha \right) - \sum_{k=1}^{\infty} \lambda^{(s-2+\alpha)k} \cos_\alpha \left(\lambda^{\alpha k} (x)^\alpha \right) $$

$$ = \sum_{k=1}^{\infty} \lambda^{(s-2+\alpha)k} \left[\cos_\alpha \left(\lambda^{\alpha k} x^\alpha \right) \cos_\alpha \left(\lambda^{\alpha k} h^\alpha \right) - \sin_\alpha \left(\lambda^{\alpha k} x^\alpha \right) \sin_\alpha \left(\lambda^{\alpha k} h^\alpha \right) \right] - \sum_{k=1}^{\infty} \lambda^{(s-2+\alpha)k} \cos_\alpha \left(\lambda^{\alpha k} (x)^\alpha \right) $$

$$ = \sum_{k=1}^{\infty} \lambda^{(s-2+\alpha)k} \left[\cos_\alpha \left(\lambda^{\alpha k} x x^\alpha \right) \left(\cos_\alpha \left(\lambda^{\alpha k} h^\alpha \right) - 1 \right) - \sin_\alpha \left(\lambda^{\alpha k} x^\alpha \right) \sin_\alpha \left(\lambda^{\alpha k} h^\alpha \right) \right] $$

From the series expansion of $\sin_\alpha \left(\lambda^{\alpha k} x^\alpha \right)$ and $\cos_\alpha \left(\lambda^{\alpha k} x^\alpha \right)$ and also from the Figure 1 andFigure 2 it is clear that for small x, $\sin_\alpha \left(\lambda^{\alpha k} x^\alpha \right) \approx \lambda^{\alpha k} x^\alpha$ and $\cos_\alpha \left(\lambda^{\alpha k} x^\alpha \right) \approx 1$ also both $\left| \sin_\alpha \left(\lambda^{\alpha k} x^\alpha \right) \right|$ and $\left| \cos_\alpha \left(\lambda^{\alpha k} x^\alpha \right) \right|$ is less than or equal to 1. Therefore, with above observation that is for small h, $\sin_\alpha \left(\lambda^{\alpha k} h^\alpha \right) \approx \lambda^{\alpha k} h^\alpha$, $\cos_\alpha \left(\lambda^{\alpha k} h^\alpha \right) - 1 \approx 0$ and for large h, $\cos_\alpha \left(\lambda^{\alpha k} h^\alpha \right) \approx 0$ we write the following

$$\left|W_{\alpha}^{(\alpha)}\left[(x+h)^{\alpha}\right]-W_{\alpha}^{(\alpha)}\left[x^{\alpha}\right]\right| \le \sum_{k=1}^{\infty}\lambda^{(s-2+\alpha)k}\left[\left|\cos_{\alpha}\left(\lambda^{\alpha k}x^{\alpha}\right)\right|\left|\cos_{\alpha}\left(\lambda^{\alpha k}h^{\alpha}\right)-1\right|+\left|\sin_{\alpha}\left(\lambda^{\alpha k}x^{\alpha}\right)\right|\left|\sin_{\alpha}\left(\lambda^{\alpha k}h^{\alpha}\right)\right|\right]$$

$$\le \sum_{k=1}^{\infty}\lambda^{(s-2+\alpha)k}\left[\min\left(\lambda^{\alpha k}h^{\alpha},1\right)\right]$$

Choose $0<h<1$ then one can find positive integer m such that $\lambda^{-(m+1)}\le h\le \lambda^{-m}$ then as per our earlier derivation for $W_{\alpha}\left(x^{\alpha}\right)$ we do the following steps

$$\left|W_{\alpha}^{(\alpha)}\left[(x+h)^{\alpha}\right]-W_{\alpha}^{(\alpha)}\left[x^{\alpha}\right]\right| \le \sum_{k=1}^{m}\lambda^{(s-2+\alpha)k}\lambda^{\alpha k}h^{\alpha}+\sum_{k=m+1}^{\infty}(1)\lambda^{(s-2+\alpha)k}$$

$$=h^{\alpha}\sum_{k=1}^{m}\lambda^{(s-2+2\alpha)}+\sum_{k=m+1}^{\infty}(1)\lambda^{(s-2+\alpha)k}$$

$$=h^{\alpha}\left(\lambda^{(s-2+2\alpha)}\frac{\lambda^{(s-2+2\alpha)m}-1}{\lambda^{(s-2+2\alpha)}-1}\right)+1\left(\frac{\lambda^{(s-2+\alpha)(m+1)}}{1-\lambda^{(s-2+\alpha)}}\right)$$

$$\le h^{\alpha}\frac{\lambda^{(s-2+2\alpha)(m+1)}}{\lambda^{(s-2+2\alpha)}-1}+\frac{\lambda^{(s-2+\alpha)(m+1)}}{1-\lambda^{(s-2+\alpha)}}$$

With $\lambda^{-(m+1)}\le h\le \lambda^{-m}$, that is $\lambda^{(m+1)}\ge h^{-1}\ge \lambda^{m}$ we get the following

$$\left|W_{\alpha}^{(\alpha)}\left[(x+h)^{\alpha}\right]-W_{\alpha}^{(\alpha)}\left[x^{\alpha}\right]\right| \le h^{\alpha}\frac{\lambda^{(s-2+2\alpha)(m+1)}}{\lambda^{(s-2+2\alpha)}-1}+\frac{\lambda^{(s-2+\alpha)(m+1)}}{1-\lambda^{(s-2+\alpha)}}$$

$$\le h^{\alpha}\frac{h^{-(s-2+2\alpha)}}{\lambda^{(s-2+2\alpha)}-1}+\frac{h^{-(s-2+\alpha)}}{1-\lambda^{(s-2+\alpha)}}$$

$$\le\left(\frac{1}{\lambda^{(s-2+2\alpha)}-1}+\frac{1}{1-\lambda^{(s-2+\alpha)}}\right)h^{2-s-\alpha}$$

$$\le C_2h^{2-s-\alpha}$$

where $C_2 = \frac{1}{\lambda^{(s-2+2\alpha)}-1} + \frac{1}{1-\lambda^{(s-2+\alpha)}}$. From definition of Holderian function and above discussion it is clear that α-order $(0<\alpha<1)$ fractional derivative of fractional Weierstrass function is also Holder continuous

with Holder exponent $2-s-\alpha$. This shows that Hausdorff dimension of graph of fractional Weierstrass function is $[2-(2-s-\alpha)] = s+\alpha$ (by lemma-1). The graph dimension increased by fractional order for fractional derivative of Weierstrass function by amount of fractional derivative-the graph becomes rougher.

5. CONCLUSION

The fractional Weierstrass function is a continuous function for all real values of the arguments, and its box dimension and Holder exponent are independent of fractional order that incorporates to the fractional Weierstrass functions. Again the Box dimension of fractional derivative of the fractional Weierstrass increases with increase of order of fractional derivative. This invariant nature of the roughness index of fractional Weierstrass function when generalized with fractional trigonometric function is remarkable. The other embodiment in similar lines as in this paper to get different fractional Weierstrass function is under development.

ACKNOWLEDGEMENTS

Acknowledgments are to Board of Research in Nuclear Science (BRNS), Department of Atomic Energy Government of India for financial assistance received through BRNS research project no. 37(3)/14/46/2014-BRNS with BSC BRNS, title "Characterization of unreachable (Holderian) functions via Local Fractional Derivative and Deviation Function". Authors are also thankful to the reviewer for his valuable comments which has helped to improve the paper.

REFERENCES

1. Mandelbrot, B.B. (1982) The Geometry of Nature. Freeman, San Francisco.

2. Peitgen, H. and Saupe, D., Eds. (1988) The Science of Fractal Images. Springer-Verlag, New York.

3. Ghosh, U. and Khan, D.K. (2014) Information, Fractal, Percolation and Geo-Environmental Complexities. LAP LAMBERT Academic Publishing.

4. Ross, B. (1977) The Development of Fractional Calculus 1695-1900. Historia Mathematica, 4, 75-89.

5. Diethelm, K. (2010) The Analysis of Fractional Differential Equations. Springer-Verlag.

6. Kilbas, A., Srivastava, H.M. and Trujillo, J.J. (2006) Theory and Applications of Fractional Differential Equations. North-Holland Mathematics Studies, Elsevier Science, Amsterdam, 1-523.

7. Miller, K.S. and Ross, B. (1993) An Introduction to the Fractional Calculus and Fractional Differential Equations. John Wiley & Sons, New York.

8. Samko, S.G., Kilbas, A.A. and Marichev, O.I. (1993) Fractional Integrals and Derivatives. Gordon and Breach Science, Yverdon.

9. Das, S. (2011) Functional Fractional Calculus. 2nd Edition, Springer-Verlag.

10. Jumarie, G. (2007) Fractional Partial Differential Equations and Modified Riemann-Liouville Derivatives. Method for Solution. Journal of Applied Mathematics and Computing, 24, 31-48.

11. Podlubny, I. (1999) Fractional Differential Equations, Mathematics in Science and Engineering. Academic Press, San Diego, 198.

12. Liang, Y.S. and Su, W. (2007) Connection between the Order of Fractional Calculus and Fractional Dimensions of a Type of Fractal Functions. Analysis in Theory and Applications, 23, 354-362.

13. Falconer, J. (1990) Fractal Geometry: Mathematical Foundations and Applications. John Wiley Sons Inc., New York.

14. Johensen, J. (2010) Simple Proofs of Nowhere-Differentiability for Weierstrass's Function and Cases of Slow Growth. Journal of Fourier Analysis and Applications, 16, 17-33.

15. Zhou, S.P., Yao, K. and Su, W.Y. (2004) Fractional Integrals of the Weierstrass Functions: The Exact Box Dimension. Analysis in Theory and Applications, 20, 332-341.

16. Kolwankar, K.M. and Gangal, A.D. (1997) Holder Exponent of Irregular Signals and Local Fractional Derivatives. Pramana, 48, 49-68.

17. Jumarie, G. (2006) Modified Riemann-Liouville Derivative and Fractional Taylor Series of Non-Differentiable Functions Further Results. Computers and Mathematics with Applications, 51, 1367-1376.

18. Jumarie, G. (2008) Fourier's Transformation of Fractional Order via Mittag-Leffler Function and Modified Riemann-Liouville Derivatives. Journal of Applied Mathematics and Informatics, 26, 1101-1121.

19. Erdelyi, A. (1954) Asymptotic Expansions. Dover Publications, New York.

20. Erdelyi, A., Ed. (1954) Tables of Integral Transforms. Volume 1, McGraw-Hill, New York.

21. Erdelyi, A. (1950) On Some Functional Transformation Univ Potitec Torino 1950.

22. Mittag-Leffler, G.M. (1903) Sur la nouvelle fonction Eα(x). Comptes Rendus de l'Académie des Sciences, 137, 554-558.

23. Hunt, B.R. (1998) The Hausdorff Dimension of Graph of Weierstrass Functions. Proceedings of the American Mathematical Society, 126, 791-801.

24. Wen, Y.Z. (2000) Mathematical Foundations of Fractal Geometry. Shanghai Science and Technology Educational Publishing House, Shanghai.

25. Zahle, M. and Ziezold, H. (1996) Fractional Derivatives of Weierstrass-Type Functions. Journal of Computational and Applied Mathematics, 76, 265-275.

6

CHAPTER

Applications of Multivalent Functions Associated with Generalized Fractional Integral Operator

Jae Ho Choi

Department of Mathematics Education, Daegu National University of Education, Daegu, South Korea

ABSTRACT

By using a method based upon the Briot-Bouquet differential subordination, we investigate some subordination properties of the generalized fractional integral operator $\mathcal{J}_{0,z}^{\lambda,\mu,\nu}$ which was defined by Owa, Saigo and Srivastava [1]. Some interesting further consequences are also considered.

Keywords: Multivalent Functions; Subordination; Gaussian Hypergeometric Function; Fractional Integral Operator

1. INTRODUCTION

Let $\mathcal{A}_n(p)$ denote the class of functions $f(z)$ of the form

$$f(z)=z^p+\sum_{k=n}^{\infty}a_{p+k}z^{p+k},(p,n\in\mathbb{N}:=\{1,2,3,\cdots\}), \quad (1.1)$$

which are analytic in the open unit disk $\mathbb{U}=\{z:z\in\mathbb{C}\text{ and }|z|<1\}$. Also let f and g be analytic in $\mathbb{U}$ with $f(0)=g(0)$. Then we say that f is subordinate to g in $\mathbb{U}$, written $f\prec g$ or $f(z)\prec g(z)$, if there exists the Schwarz function w, analytic in $\mathbb{U}$ such that $w(0)=0$, $|w(z)|<1$ and $f(z)=g(w(z))(z\in\mathbb{U})$. We also observe that

$f(z)\prec g(z)$ in $\mathbb{U}$

if and only if

$$f(0)=g(0)\text{ and }f(\mathbb{U})\subset g(\mathbb{U})$$

whenever g is univalent in $\mathbb{U}$.

Let a, b and c be complex numbers with $c\neq 0,-1,-2,\cdots$. Then the Gaussian/classical hypergeometric function ${}_2F_1(a,b;c;z)$ is defined by

$$ {}_2F_1(a,b;c;z)=\sum_{k=0}^{\infty}\frac{(a)_k(b)_k}{(c)_k}\frac{z^k}{k!}, \tag{1.2}$$

where $(\eta)_k$ is the Pochhammer symbol defined, in terms of the Gamma function, by

$$(\eta)_k=\frac{\Gamma(\eta+k)}{\Gamma(\eta)}=\begin{cases}1, & (k=0)\\ \eta(\eta+1)\cdots(\eta+k-1), & (k\in\mathbb{N}).\end{cases} \tag{1.3}$$

The hypergeometric function ${}_2F_1(a,b;c;z)$ is analytic in $\mathbb{U}$ and if a or b is a negative integer, then it reduces to a polynomial.

For each A and B such that $-1\le B<A\le 1$, let us define the function

$$h(A,B;z)=\frac{1+Az}{1+Bz},(z\in\mathbb{U}). \tag{1.4}$$

It is well known that $h(A,B;z)$, for $-1\le B\le 1$, is the conformal map of the unit disk onto the disk symmetrical respect to the real axis having the center $(1-AB)/(1-B^2)$ and the radius $(A-B)/(1-B^2)$. The boundary circle cuts the real axis at the points $(1-A)/(1-B)$ and $(1+A)/(1+B)$.

Many essentially equivalent definitions of fractional calculus have been given in the literature (cf., e.g. [2,3]). We state here the following definition due to Saigo [4] (see also [1,5]).

Definition 1. For $\lambda > 0$, $\mu, \nu \in \mathbb{R}$, the fractional integral operator $\mathcal{I}_{0,z}^{\lambda,\mu,\nu}$ is defined by

$$\mathcal{I}_{0,z}^{\lambda,\mu,\nu} f(z) = \frac{z^{-\lambda-\mu}}{\Gamma(\lambda)} \int_0^z (z-\zeta)^{\lambda-1}\, {}_2F_1\left(\lambda+\mu, -\nu; \lambda; 1-\frac{\zeta}{z}\right) f(\zeta)\,\mathrm{d}\zeta, \tag{1.5}$$

where ${}_2F_1$ is the Gaussian hypergeometric function defined by (1.2) and $f(z)$ is taken to be an analytic function in a simply-connected region of the z-plane containing the origin with the order

$$f(z) = \mathcal{O}\left(|z|^{\epsilon}\right)(z \to 0)$$

for $\epsilon > \max\{0, \mu-\nu\} - 1$, and the multiplicity of $(z-\zeta)^{\lambda-1}$ is removed by requiring that $\log(z-\zeta)$ to be real when $z-\zeta > 0$.

The definition (1.5) is an interesting extension of both the Riemann-Liouville and Erdélyi-Kober fractional operators in terms of Gauss's hypergeometric functions.

With the aid of the above definition, Owa, Saigo and Srivastava [1] defined a modification of the fractional integral operator $\mathcal{J}_{0,z}^{\lambda,\mu,\nu}$ by

$$\mathcal{J}_{0,z}^{\lambda,\mu,\nu} f(z) = \frac{\Gamma(p+1-\mu)\Gamma(\lambda+p+1+\nu)}{\Gamma(p+1)\Gamma(p+1-\mu+\nu)} z^{\mu} \mathcal{I}_{0,z}^{\lambda,\mu,\nu} f(z) \tag{1.6}$$

for $f(z) \in \mathcal{A}_n(p)$ and $\mu - \nu - p < 1$. Then it is observed that $\mathcal{J}_{0,z}^{\lambda,\mu,\nu}$ also maps $\mathcal{A}_n(p)$ onto itself as follows:

$$\mathcal{J}_{0,z}^{\lambda,\mu,\nu} f(z) = z^p + \sum_{k=n}^{\infty} \frac{(p+1)_k (p+1-\mu+\nu)_k}{(p+1-\mu)_k (\lambda+p+1+\nu)_k} a_{p+k} z^{p+k}, \quad (\lambda > 1; \mu - \nu - p < 1; f \in \mathcal{A}_n(p)). \tag{1.7}$$

We note that $\mathcal{J}_{0,z}^{\alpha,0,\beta-1} f(z) = \mathcal{O}_{\beta}^{\alpha} f(z), (\alpha \geq 0; \beta > -1)$, where the operator $\mathcal{O}_{\beta}^{\alpha}$ was introduced and studied by Jung, Kim and Srivastava [6] (see also [7]).

It is easily verified from (1.7) that

$$z\left(\mathcal{J}_{0,z}^{\lambda,\mu,\nu} f(z)\right)' = (\lambda+\nu+p)\mathcal{J}_{0,z}^{\lambda-1,\mu,\nu} f(z) - (\lambda+\nu)\mathcal{J}_{0,z}^{\lambda,\mu,\nu} f(z). \tag{1.8}$$

The identity (1.8) plays an important and significant role in obtaining our results.

Recently, by using the general theory of differential subordination, several authors (see, e.g. [7-9]) considered some interesting properties of multivalent functions associated with various integral operators. In this manuscript, we shall derive some subordination properties of the fractional integral operator $\mathcal{J}_{0,z}^{\lambda,\mu,\nu}$ by using the technique of differential subordination.

2. MAIN RESULTS

In order to establish our results, we shall need the following lemma due to Miller and Mocanu [10].

Lemma 1. Let $h(t)$ be analytic and convex univalent in $\mathbb{U}$ with $h(0)=1$, and let $g(z)=1+b_n z^n+b_{n+1}z^{n+1}+\cdots$ be analytic in $\mathbb{U}$. If

$$g(z)+\frac{1}{c}zg'(z)\prec h(z), \tag{2.1}$$

then for $c\neq 0$ and $\operatorname{Re} c\geq 0$,

$$g(z)\prec \frac{c}{n}z^{-c/n}\int_0^z t^{c/n-1}h(t)\,\mathrm{d}t. \tag{2.2}$$

We begin by proving the following theorem.

Theorem 1. Let $-1\leq B<A\leq 1$, $\lambda>1$, $\lambda+\nu>-p$, $\mu-\nu-p<1$, $\mu-1<p$ and $0<\alpha<1$, and let $f(z)=z^p+\sum_{k=n}^{\infty}a_{p+k}z^{p+k}\in\mathcal{A}_n(p)$. Suppose that

$$\sum_{k=n}^{\infty} c_k \left|a_{p+k}\right| \leq 1, \tag{2.3}$$

where

$$c_k = \frac{1-B}{A-B} \frac{\left[\lambda+p+\nu+k(1-\alpha)\right](p+1)_k (p+1-\mu+\nu)_k}{(\lambda+p+\nu)(p+1-\mu)_k (\lambda+p+1+\nu)_k} \tag{2.4}$$

and $(\eta)_k$ is given by (1.3).

1) If $-1 \leq B < 0$, then

$$(1-\alpha)\frac{\mathcal{J}_{0,z}^{\lambda-1,\mu,\nu} f(z)}{z^p} + \alpha \frac{\mathcal{J}_{0,z}^{\lambda,\mu,\nu} f(z)}{z^p} \prec h(A,B;z). \tag{2.5}$$

2) If $-1 \leq B < 0$ and $\gamma \geq 1$, then

$$\operatorname{Re}\left\{\left(\frac{\mathcal{J}_{0,z}^{\lambda,\mu,\nu} f(z)}{z^p}\right)^{1/\gamma}\right\}$$

$$> \left\{\frac{\lambda+\nu+p}{n(1-\alpha)} \int_0^1 u^{\frac{\lambda+\nu+p}{n(1-\alpha)}-1} \left(\frac{1-Au}{1-Bu}\right) \mathrm{d}u\right\}^{1/\gamma}, (z \in U). \tag{2.6}$$

The result is sharp.

Proof. 1) If we set

$$L=(1-\alpha)\frac{\mathcal{J}_{0,z}^{\lambda-1,\mu,\nu}f(z)}{z^p}+\alpha\frac{\mathcal{J}_{0,z}^{\lambda,\mu,\nu}f(z)}{z^p},$$

then, from (1.7) we see that

$$L=1+\sum_{k=n}^{\infty}\frac{[\lambda+p+\nu+k(1-\alpha)](p+1)_k(p+1-\mu+\nu)_k}{(\lambda+p+\nu)(p+1-\mu)_k(\lambda+p+1+\nu)_k}a_{p+k}z^k. \tag{2.7}$$

For $-1\le B<0$ and $z\in\mathbb{U}$, it follows from (2.3) that

$$\left|\frac{L-1}{A-BL}\right|=\left|\frac{\sum_{k=n}^{\infty}\frac{[\lambda+p+\nu+k(1-\alpha)](p+1)_k(p+1-\mu+\nu)_k}{(\lambda+p+\nu)(p+1-\mu)_k(\lambda+p+1+\nu)_k}a_{p+k}z^k}{A-B-B\sum_{k=n}^{\infty}\frac{[\lambda+p+\nu+k(1-\alpha)](p+1)_k(p+1-\mu+\nu)_k}{(\lambda+p+\nu)(p+1-\mu)_k(\lambda+p+1+\nu)_k}a_{p+k}z^k}\right|\le\frac{\sum_{k=n}^{\infty}c_k|a_{p+k}|}{1-B+B\sum_{k=n}^{\infty}c_k|a_{p+k}|}\le1, \tag{2.8}$$

which implies that

$$(1-\alpha)\frac{\mathcal{J}_{0,z}^{\lambda-1,\mu,\nu}f(z)}{z^p}+\alpha\frac{\mathcal{J}_{0,z}^{\lambda,\mu,\nu}f(z)}{z^p}\prec h(A,B;z).$$

2) Let

$$g(z)=\frac{\mathcal{J}_{0,z}^{\lambda,\mu,\nu}f(z)}{z^p},\ (f\in\mathcal{A}_n(p)). \tag{2.9}$$

Then the function $g(z)=1+b_nz^n+b_{n+1}z^{n+1}+\cdots$ is analytic in $\mathbb{U}$. Using (1.8) and (2.9), we have

$$\frac{\mathcal{J}_{0,z}^{\lambda-1,\mu,\nu} f(z)}{z^p} = g(z) + \frac{1}{\lambda+\nu+p} zg'(z). \tag{2.10}$$

From (2.5), (2.9) and (2.10) we obtain

$$(1-\alpha)\frac{\mathcal{J}_{0,z}^{\lambda-1,\mu,\nu} f(z)}{z^p} + \alpha \frac{\mathcal{J}_{0,z}^{\lambda,\mu,\nu} f(z)}{z^p}$$
$$= g(z) + \frac{1-\alpha}{\lambda+\nu+p} zg'(z) \prec h(A,B;z).$$

Thus, by applying Lemma 1, we observe that

$$g(z) \prec \frac{\lambda+\nu+p}{n(1-\alpha)} z^{-\frac{\lambda+\nu+p}{n(1-\alpha)}} \int_0^z t^{\frac{\lambda+\nu+p}{n(1-\alpha)}-1} \left(\frac{1+At}{1+Bt}\right) \mathrm{d}t$$

or

$$\frac{\mathcal{J}_{0,z}^{\lambda,\mu,\nu} f(z)}{z^p} = \frac{\lambda+\nu+p}{n(1-\alpha)} \int_0^1 u^{\frac{\lambda+\nu+p}{n(1-\alpha)}-1} \left(\frac{1+Auw(z)}{1+Buw(z)}\right) du, \tag{2.11}$$

where $w(z)$ is analytic in $\mathbb{U}$ with $w(0)=0$ and $|w(z)|<1 \ (z \in \mathbb{U})$. In view of $-1 \le B < A \le 1$ and $\lambda+\nu > -p$, we conclude from (2.11) that

$$\operatorname{Re}\left\{\frac{\mathcal{J}_{0,z}^{\lambda,\mu,\nu} f(z)}{z^p}\right\} > \frac{\lambda+\nu+p}{n(1-\alpha)}\int_0^1 u^{\frac{\lambda+\nu+p}{n(1-\alpha)}-1}\left(\frac{1-Au}{1-Bu}\right)du,$$

$(z \in \mathbb{U})$.

(2.12)

Since $\operatorname{Re}\left(w^{1/\gamma}\right) \geq (\operatorname{Re} w)^{1/\gamma}$ for $\operatorname{Re} w > 0$ and $\gamma \geq 1$, from (2.12) we see that the inequality (2.6) holds.

To prove sharpness, we take $f(z) \in \mathcal{A}_n(p)$ defined by

$$\frac{\mathcal{J}_{0,z}^{\lambda,\mu,\nu} f(z)}{z^p} = \frac{\lambda+\nu+p}{n(1-\alpha)}\int_0^1 u^{\frac{\lambda+\nu+p}{n(1-\alpha)}-1}\left(\frac{1+Auz^n}{1+Buz^n}\right)du.$$

For this function we find that

$$(1-\alpha)\frac{\mathcal{J}_{0,z}^{\lambda-1,\mu,\nu} f(z)}{z^p} + \alpha\frac{\mathcal{J}_{0,z}^{\lambda,\mu,\nu} f(z)}{z^p} = \frac{1+Az^n}{1+Bz^n}$$

and

$$\frac{\mathcal{J}_{0,z}^{\lambda,\mu,\nu} f(z)}{z^p} \to \frac{\lambda+\nu+p}{n(1-\alpha)}\int_0^1 u^{\frac{\lambda+\nu+p}{n(1-\alpha)}-1}\frac{1-Au}{1-Bu}du \text{ as } z \to e^{i\pi/n}.$$

Hence the proof of Theorem 1 is evidently completed.

Theorem 2. Let $-1 \leq B < A \leq 1$, $\lambda > 1$, $\lambda+\nu > -p$, $\mu-\nu-p < 1$, $\mu-1 < p$ and $0 < \alpha < 1$. Suppose that

$f(z)=z^p+\sum_{k=n}^{\infty}a_{p+k}z^{p+k}\in\mathcal{A}_n(p)$, $s_1(z)=z^p$

and $s_m(z)=z^p+\sum_{k=n}^{n+m-2}a_{p+k}z^{p+k}\ (m\geq 2)$. If the sequence $\{c_k\}$ is nondecreasing with

$$c_k\geq\frac{(1-B)\left[\lambda+p+v+k(1-\alpha)\right]}{(A-B)(\lambda+p+v)}(k\geq n), \tag{2.13}$$

where c_k is given by (2.4) and satisfies the condition (2.3), then

$$\operatorname{Re}\left\{\frac{\mathcal{J}_{0,z}^{\lambda,\mu,v}f(z)}{s_m(z)}\right\}>0 \tag{2.14}$$

and

$$\operatorname{Re}\left\{\frac{s_m(z)}{\mathcal{J}_{0,z}^{\lambda,\mu,v}f(z)}\right\}>0. \tag{2.15}$$

Each of the bounds in (2.14) and (2.15) is best possible for $m\in\mathbb{N}$.

Proof. We prove the bound in (2.14). The bound in (2.15) is immediately obtained from (2.14) and will be omitted. Let

$$h(z)=\frac{\mathcal{J}_{0,z}^{\lambda,\mu,v}f(z)}{s_m(z)}\left(f\in\mathcal{A}_n(p);z\in\mathbb{U}\right).$$

Then, from (1.7) we observe that

$$h(z)=1+\frac{\sum_{k=n}^{n+m-2}(\delta_k-1)a_{p+k}z^k+\sum_{k=n+m-1}^{\infty}\delta_k a_{p+k}z^k}{1+\sum_{k=n}^{n+m-2}a_{p+k}z^k},$$

where, for convenience,

$$\delta_k=\frac{(p+1)_k(p+1-\mu+\nu)_k}{(p+1-\mu)_k(\lambda+p+1+\nu)_k}.$$

It is easily seen from (2.4) and (2.13) that $c_k>1$ and

$$\delta_k=\frac{(A-B)(\lambda+p+\nu)}{(1-B)[\lambda+p+\nu+k(1-\alpha)]}c_k\geq 1. \tag{2.16}$$

Hence, by applying (2.3) and (2.16), we have

$$\left|\frac{h(z)-1}{h(z)+1}\right|=\left|\frac{\sum_{k=n}^{n+m-2}(\delta_k-1)a_{p+k}z^k+\sum_{k=n+m-1}^{\infty}\delta_k a_{p+k}z^k}{2+\sum_{k=n}^{n+m-2}(\delta_k+1)a_{p+k}z^k+\sum_{k=n+m-1}^{\infty}\delta_k a_{p+k}z^k}\right|$$

$$\leq\frac{\sum_{k=n}^{n+m-2}(\delta_k-1)|a_{p+k}|+\sum_{k=n+m-1}^{\infty}\delta_k|a_{p+k}|}{2-\sum_{k=n}^{n+m-2}(\delta_k+1)|a_{p+k}|-\sum_{k=n+m-1}^{\infty}\delta_k|a_{p+k}|}\leq 1\,(z\in\mathbb{U})$$

which readily yields the inequality (2.14).

If we take $f(z)=z^{p}-z^{p+n+m-1}$, then

$$\frac{f(z)}{s_m(z)}=1-z^{n+m-1}\to 0 \text{ as } z\to 1^{-}.$$

This show that the bound in (2.14) is best possible for each m, which proves Theorem 2.

Finally, we consider the generalized Bernardi-LiveraLivingston integral operator $\mathcal{L}_{\sigma}(\sigma>-p)$ defined by (cf. [11-13])

$$\mathcal{L}_{\sigma}(f)(z):=\frac{\sigma+p}{z^{\sigma}}\int_{0}^{z}t^{\sigma-1}f(t)\mathrm{d}t\left(f\in\mathcal{A}_n(p);\sigma>-p\right). \tag{2.17}$$

Theorem 3. Let $-1\le B<A\le 1$, $\sigma>-p$, $\lambda>1$, $\lambda+\nu>-p$, $\mu-\nu-p<1$, $\mu-1<p$ and $0<\alpha<1$ and let $f(z)=z^{p}+\sum_{k=n}^{\infty}a_{p+k}z^{p+k}\in\mathcal{A}_n(p)$. Suppose that

$$\sum_{k=n}^{\infty}d_k\left|a_{p+k}\right|\le 1, \tag{2.18}$$

where

$$d_k=\frac{1-B}{A-B}\frac{\left[\sigma+p+k(1-\alpha)\right](p+1)_k(p+1-\mu+\nu)_k}{(\sigma+p+k)(p+1-\mu)_k(\lambda+p+1+\nu)_k}$$

and $(\eta)_k$ is given by (1.3).

1) If $-1 \le B < 0$, then

$$(1-\alpha)\frac{\mathcal{J}_{0,z}^{\lambda,\mu,\nu} f(z)}{z^p} + \alpha \frac{\mathcal{J}_{0,z}^{\lambda,\mu,\nu} \mathcal{L}_\sigma(f)(z)}{z^p} \prec h(A,B;z). \tag{2.19}$$

2) If $-1 \le B < 0$ and $\gamma \ge 1$, then

$$\operatorname{Re}\left\{\left(\frac{\mathcal{J}_{0,z}^{\lambda,\mu,\nu} \mathcal{L}_\sigma(f)(z)}{z^p}\right)^{1/\gamma}\right\}$$

$$> \left\{\frac{\sigma+p}{n(1-\alpha)}\int_0^1 u^{\frac{\sigma+p}{n(1-\alpha)}-1}\left(\frac{1-Au}{1-Bu}\right)du\right\}^{1/\gamma} \quad (z \in U). \tag{2.20}$$

The result is sharp.

Proof. 1) If we put

$$M = (1-\alpha)\frac{\mathcal{J}_{0,z}^{\lambda,\mu,\nu} f(z)}{z^p} + \alpha \frac{\mathcal{J}_{0,z}^{\lambda,\mu,\nu} \mathcal{L}_\sigma(f)(z)}{z^p},$$

then, from (1.7) and (2.17) we have

$$M = 1 + \sum_{k=n}^{\infty} \frac{[\sigma+p+k(1-\alpha)](p+1)_k(p+1-\mu+\nu)_k}{(\sigma+p+k)(p+1-\mu)_k(\lambda+p+1+\nu)_k}$$
$$\cdot a_{p+k} z^k.$$

Therefore, by using same techniques as in the proof of Theorem 1 1), we obtain the desired result.

2) From (2.17) we have

$$(\sigma+p)\mathcal{J}_{0,z}^{\lambda,\mu,\nu}f(z)$$
$$=\sigma\mathcal{J}_{0,z}^{\lambda,\mu,\nu}\mathcal{L}_{\sigma}(f)(z)+z\left(\mathcal{J}_{0,z}^{\lambda,\mu,\nu}\mathcal{L}_{\sigma}(f)(z)\right)'. \quad (2.21)$$

Let

$$g(z)=\frac{\mathcal{J}_{0,z}^{\lambda,\mu,\nu}\mathcal{L}_{\sigma}(f)(z)}{z^{p}}(z\in\mathbb{U}). \quad (2.22)$$

Then, by virtue of (2.21), (2.22) and (2.19), we observe that

$$(1-\gamma)\frac{\mathcal{J}_{0,z}^{\lambda,\mu,\nu}f(z)}{z^{p}}+\gamma\frac{\mathcal{J}_{0,z}^{\lambda,\mu,\nu}\mathcal{L}_{\sigma}(f)(z)}{z^{p}}$$
$$=g(z)+\frac{1-\gamma}{\sigma+p}zg'(z)\prec h(A,B;z).$$

Hence, by applying the same argument as in the proof of Theorem 1 2), we obtain (2.20), which evidently proves Theorem 3.

3. ACKNOWLEDGEMENTS

This work was supported by Daegu National University of Education Research grant in 2011.

REFERENCES

1. S. Owa, M. Saigo and H. M. Srivastava, "Some Characterization Theorems for Starlike and Convex Functions Involving a Certain Fractional Integral Operator," Journal of Mathematical Analysis and Applications, Vol. 140, No. 2, 1989, pp. 419-426.

2. S. G. Samko, A. A. Kilbas and O. I. Marichev, "Fractional Integral and Derivatives, Theory and Applications," Gordon and Breach, New York, Philadelphia, London, Paris, Montreux, Toronto, Melbourne, 1993.

3. H. M. Srivastava and R. G. Buschman, "Theory and Applications of Convolution Integral Equations," Kluwer Academic Publishers, Dordrecht, Boston, London, 1992.

4. M. Saigo, "A Remark on Integral Operators Involving the Gauss Hypergeometric Functions," Mathematical Reports, Kyushu University, Vol. 11, No. 2, 1977-1978, pp. 135- 143.

5. J. H. Choi, "Note on Differential Subordination Associated with Fractional Integral Operator," Far East Journal of Mathematical Sciences, Vol. 26, No. 2, 2007, pp. 499- 511.

6. I. B. Jung, Y. C. Kim and H. M. srivastava, "The Hardy Space of Analytic Functions Associated with Certain OneParameter Families of Integral Operators," Journal of Mathematical Analysis and Applications, Vol. 176, No. 1, 1993, pp. 138-147.

7. J.-L. Liu, "Notes on Jung-Kim-Srivastava Integral Operator," Journal of Mathematical Analysis and Applications, Vol. 294, No. 1, 2004, pp. 96-103.

8. R. M. EL-Ashwash and M. K. Aouf, "Some Subclasses of Multivalent Functions Involving the Extended Fractional Differintegral Operator," Journal of Mathematical Inequalities, Vol. 4, No. 1, 2010, pp. 77-93.

9. J. Patel, A. K. Mishra and H. M. Srivastava, "Classes of Multinalent Analytic Functions Involving the DziokSrivastava Operator," Computers and Mathematics with Applications, Vol. 54, No. 5, 2007, pp. 599-616.

10. S. S. Miller and P. T. Mocanu, "Differential Subordinations and Univalent Functions," Michigan Mathematical Journal, Vol. 28, No. 2, 1981, pp. 157-172.

11. S. D. Bernardi, "Convex and Starlike Univalent Functions," Transactions of the American Mathematical Society, Vol. 135, 1969, pp. 429-446.

12. R. J. Libera, "Some Classes of Regular Univalent Functions," Proceedings of the American Mathematical Society, Vol. 16, No. 4, 1965, pp. 755-758.

13. H. M. Srivastava and S. Owa, Eds., "Current Topics in Analytic Function Theory," World Scientific Publishing Company, Singapore, New Jersey, London, Hong Kong, 1992.

7

CHAPTER

On Humbert Matrix Polynomials of Two Variables

Ghazi S. Khammash[1], A. Shehata[2]

[1]*Department of Mathematics, Al-Aqsa University, Gaza Strip, Palestine*

[2]*Department of Mathematics, Faculty of Science, Assiut University, Assiut, Egypt*

ABSTRACT

In this paper we introduce Humbert matrix polynomials of two variables. Some hypergeometric matrix representations of the Humbert matrix polynomials of two variables, the double generating matrix functions and expansions of the Humbert matrix polynomials of two variables in series of Hermite polynomials are given. Results of Gegenbauer matrix polynomials of two variables follow as particular cases of Humbert matrix polynomials of two variables.

Keywords: Humbert Matrix Polynomials of Two Variables; Hypergeometric Matrix Function; Matrix Functional Calculus

1. INTRODUCTION

The special matrix functions appear in statistics, lie group theory and number theory [1-4] and the matrix polynomials have become more important and some results in the theory of classical orthogonal polynomials have been extended to orthogonal matrix polynomials see for instance [5-9].

If D_0 is the complex plane cut along the negative, real axis and $\log(z)$ denotes the principal logarithm of z (Saks, S. and A. Zygmund, [10]), then $z^{1/2}$ represents $\exp(1/2\log)(z)$ if A is a matrix in $C^{N\times N}$ the set of all the eigenvalues of is denoted by the set of all the eigenvalues of A is denoted by $\sigma(A)$. If $f(z)$ and $g(z)$ are holomorphic functions of the complex variable z, which are defined in an open set Ω of the complex plane, and A is a matrix in $C^{N\times N}$ such that $\sigma(A)\subset\Omega$. Then from the properties of the matrix functional calculus, (Dunford N. and J. Schwartz J. [11]), it follows that $f(A)g(A)=g(A)f(A)$. If A is a matrix with $\sigma(A)\subset D_0$, then $A^{1/2}=\sqrt{A}$ denotes the a image by $z^{1/2}$ of the matrix functional calculus acting on the matrix A. we say that A is a positive stable matrix if $\operatorname{Re}(z)>0$ for all $z\in\sigma(A)$.

For any matrix P in $C^{N\times N}$ we will exploit the following relation due to [12]

$$(1-x)^{-p}=\sum_{n=0}^{\infty}\frac{(p)_n x^n}{n!},\ |x|<1 \tag{1}$$

Khammash [12], define the Gegenbauer matrix polynomials of two variables by

$$C_{n,k}^{A}(x,y)=\sum_{r=0}^{\left[\frac{n}{2}\right]}\sum_{j=0}^{\left[\frac{k}{2}\right]}\frac{(-1)^{r+j}(A)_{n+k-r-j}(2x)^{n-2r}(2y)^{k-2j}}{r!j!(n-2r)!(k-2j)!} \tag{2}$$

From which it follows that $C_{n,k}^{A}(x,y)$ is a matrix polynomial in two variables x and y of degree precisely n in x and k in y.

Also we recall that if $A(k,n)$ are matrix in $C^{N\times N}$ for $n\geq 0$ and $k\geq 0$ that it follows that (Defez and Jódar [14])

$$\sum_{n=0}^{\infty}\sum_{k=0}^{\infty}A(k,n)=\sum_{n=0}^{\infty}\sum_{k=0}^{n}A(k,n-k) \tag{3}$$

$$\sum_{n=0}^{\infty}\sum_{k=0}^{\infty}A(k,n)=\sum_{n=0}^{\infty}\sum_{k=0}^{[n/2]}A(k,n-k) \tag{4}$$

and, for m is a positive integer such that $n>m$, then

$$\sum_{n=0}^{\infty}\sum_{k=0}^{\infty}A(k,n)=\sum_{n=0}^{\infty}\sum_{k=0}^{[n/m]}A(k,n-k) \tag{5}$$

$$\sum_{n=0}^{\infty}\sum_{k=0}^{n}A(k,n)=\sum_{n=0}^{\infty}\sum_{k=0}^{[n/m]}A(k,n-mk+k) \tag{6}$$

We define Humbert matrix polynomials of two variables and discuss its special cases. Some hypergeometric matrix representations of the Humbert matrix polynomials of two variables, the double generating matrix functions and expansions of the Humbert matrix polynomials of two variables in series of Hermite polynomials are given. Some particular cases are also discussed.

2. DEFINITION OF HUMBERT MATRIX POLYNOMIALS OF TWO VARIABLES

Let A be a positive stable matrix in $C^{N\times N}$ for a positive integer m, we define Humbert matrix polynomials by

$$\left(1-\left(mxt-t^{m}\right)-\left(mys-s^{m}\right)\right)^{-A} = \sum_{n=0}^{\infty}\sum_{k=0}^{\infty} P_{n,k,m}(x,y,A)t^{n}s^{k} \tag{7}$$

Now (7) it can be written in the form

$$\left(1-\left(mxt-t^{m}\right)-\left(mys-s^{m}\right)\right)^{-A} = \sum_{n=0}^{\infty}\sum_{k=0}^{\infty} \frac{(A)_{n+k}\left(mxt-t^{m}\right)^{n}\left(mys-s^{m}\right)^{k}}{n!k!}$$

and by (3) and (6) respectively, one gets

$$=\sum_{n=0}^{\infty}\sum_{k=0}^{\infty}\sum_{r=0}^{n}\sum_{j=0}^{k}\frac{(A)_{n+k}(mx)^{n-r}(my)^{k-j}(-1)^{r+j}}{r!j!(n-r)!(k-j)!}$$

$$\times t^{n+(m-1)r}s^{k+(m-1)j}$$

$$=\sum_{n=0}^{\infty}\sum_{k=0}^{\infty}\sum_{r=0}^{[n/m]}\sum_{j=0}^{[k/m]}\frac{(A)_{(n+k)-(m-l)(r+j)}(-1)^{r+j}}{r!j!(n-mr)!(k-mj)!}$$

$$\times(mx)^{n-mr}(my)^{k-mj}t^{n}s^{k}$$

$$\sum_{n=0}^{\infty}\sum_{k=0}^{\infty}P_{n,k,m}(x,y,A)s^{n}t^{k}$$

$$=\sum_{n=0}^{\infty}\sum_{k=0}^{\infty}\sum_{r=0}^{[n/m]}\sum_{j=0}^{[k/m]}\frac{(A)_{(n+k)-(m-l)(r+j)}(-1)^{r+j}}{r!j!(n-mr)!(k-mj)!}$$

$$\times(mx)^{n-r}(my)^{k-j}t^{n}s^{k} \tag{8}$$

By equating the coefficients of $t^n s^k$ in (7) and (8), we obtain an explicit representation of the Humbert matrix polynomials of two variables. In the form

$$=\sum_{r=0}^{[n/m]}\sum_{j=0}^{[k/m]}\frac{(A)_{(n+k)-(m-l)(r+j)}(-1)^{r+j}\times(mx)^{n-mr}(my)^{k-mj}}{r!j!(n-mr)!(k-mj)!} \tag{9}$$

from which it follows that $P_{n,k,m}(x,y,A)$ is a matrix polynomial in two variables x and y of degree precisely n in x and k in y. In (9) setting m = 2, we get the Gegenbauer matrix polynomials of two variables [13] as particular case of the Humbert matrix polynomials of two variables.

3. HYPERGEOMETRIC MATRIX REPRESENTATION FOR $P_{n,k,m}(x,y,A)$

We study here the representation of the hypergeometric matrix representation for the Humbert matrix polynomials of two variables. There are some facts and notations used throughout the development in Sections 3 - 5, which are listed here.

Fact 1. [15] The reciprocal scalar Gamma Function $\Gamma^{-1}(z)=1/\Gamma(z)$, is an entire functions of the complex variable z. Thus, for $c \in C^{N\times N}$, the Riesz-Dunford functional calculus [11] shows that $\Gamma^{-1}(z)$ is well defined and is indeed, the inverse of $\Gamma(z)$, Hence: if $c \in C^{rxr}$ is such that $C+nI$ is invertible for every integer $n \geq 0$. Then

$$(c)_n = \Gamma(c+nI)\Gamma^{-1}(z).$$

Fact 2. [12] If A, B and C are members of C^{rxr} for which $C+nI$ is invertible for every integer $n \geq 0$. The hypergeometric matrix function $F(A,B;C;z)$ is defined by

$$F(A,B;C;z)=\sum_{n=0}^{\infty}\frac{1}{n!}\left((A)_n(B)_n\left[(C)_n\right]^{-1}\right)z^n$$

it converges for $|z| < 1$.

Notation 1. [16] For $A \in C^{r \times r}$, the matrix version of the pochhammer symbol (the shifted factorial) is

$$(A)_n = A(A+I)(A+2I)\cdots(A+(n-1)I),\quad n \geq 1 \tag{10}$$

with $(A)_0 = I$.

Note that $A = -jI$, where j is a positive integer, then $(A)_n = 0$ when ever $n > j$. Also, the product in (10) is commutative, and then it is easy to see that

$$(A)_{n+k}(A)_n(A+nI)_k$$

$$(A)_{n-k} = (-1)^k (A)_n [(I-A-nI)]^{k-1}$$

and

$$(A)_{mn} = m^m \prod_{s=1}^{m} \left(\frac{1}{m}(A+(s-1))I \right)_n$$

where m is a positive integer.

Notation 2. [17]

$$\frac{1}{(n-mk)} = \frac{(-1)^{mk}}{n!}(-n)_m k;\ 0 \leq mk \leq n$$

$$(-nI)_{mk} = m^{mk}\prod_{s=1}^{m}\left(\frac{1}{m}(s-n-1)I\right)_k$$

Now, in view of Notation 2, the explicit representation (9) for $m \geq 2$, becomes

$$P_{n,k,m}(x,y,A) = \sum_{r=0}^{\left[\frac{n}{m}\right]}\sum_{j=0}^{\left[\frac{k}{m}\right]}\frac{(A)_{(n+k)-(m-1)(r+j)}}{r!j!n!k!}\times(-1)^{(m+1)(r+j)}(mx)^{n-mr}(my)^{k-mj}$$
$$\times m^{mr}\prod_{s=1}^{m}\left(\frac{1}{m}(s-n-1)I\right)_r m^{mj}\prod_{l=1}^{m}\left(\frac{1}{m}(l-k-1)I\right)_j$$
$$=\frac{(A)_{n+k}(mx)^n(my)^k}{n!k!}\times\sum_{r=0}^{\left[\frac{n}{m}\right]}\sum_{j=0}^{\left[\frac{k}{m}\right]}\frac{(1)}{r!j!}\left[\prod_{s=1}^{m-1}\left(\frac{(-A-(n+k)I+sI)}{(m-1)}\right)_{r+j}\right]^{-1}$$
$$\times\prod_{s=1}^{m}\left(\frac{s-n-1}{m}I\right)\prod_{l=1}^{m}\left(\frac{l-k-1}{m}I\right)\times\left(\frac{1}{(m-1)^{m-1}x^m}\right)^r\left(\frac{1}{(m-1)^{m-1}y^m}\right)^j$$

Thus we get the following hypergeometric representation of Humbert matrix polynomials of two variables.

$$P_{n,k,m}(x,y,A)=\frac{(A)_{n+k}(mx)^n(my)^k}{n!k!}\times{}_mF_{m-1}^{(2)}\left[\frac{-n}{m}I,\frac{n-1}{m}I,\dots,\frac{n-m+1}{m}I;\frac{-k}{m}I,\frac{k-1}{m}I,\dots,\frac{k-m+1}{m}I;\right.$$
$$\left.\frac{-A-((n+k)-1)I}{m-1},\frac{-A-((n+k)-2)I}{m-1},\dots,\frac{-A-((n+k)-(m-1))I}{m-1};\frac{1}{(m-1)^{m-1}x^m},\frac{1}{(m-1)^{m-1}y^m}\right] \tag{11}$$

For m = 2 (11), we gives hypergeometric representation of Gegenbauer matrix polynomials of two variables [13].

The above facts and notations will be used throughout the next two sections.

4. ADDITIONAL DOUBLE GENERATING MATRIX FUNCTIONS

Now, since

$$P_{n,k,m}(x,y,A)=\sum_{r=0}^{\left[\frac{n}{m}\right]}\sum_{j=0}^{\left[\frac{k}{m}\right]}\frac{(A)_{(n+k)-(m-I)(r+j)}}{r!\,j!\,(n-mr)!\,(k-mj)!}\times(-1)^{r+j}(mx)^{n-r}(my)^{k-j}\sum_{n=0}^{\infty}\sum_{k=0}^{\infty}\frac{P_{n,k,m}(x,y,A)s^{n}t^{k}}{(A)_{n+k}}$$

$$=\sum_{n=0}^{\infty}\sum_{k=0}^{\infty}\sum_{r=0}^{\left[\frac{n}{m}\right]}\sum_{j=0}^{\left[\frac{k}{m}\right]}\frac{(A)_{(n+k)-(m-I)(r+j)}}{(A)_{n+k}\,r!\,j!\,(n-mr)!\,(k-mj)!}\times(-1)^{r+j}(mx)^{n-r}(my)^{k-j}\,t^{k+mj}s^{n+mr}$$

(12)

By using (5), one gets

$$=\sum_{n=0}^{\infty}\sum_{k=0}^{\infty}\sum_{r=0}^{\infty}\sum_{j=0}^{\infty}\frac{(A)_{(n+k)+(r+j)}}{(A)_{(n+mr)+(k+mj)}\,r!\,j!\,n!\,k!}\times(-1)^{r+j}(mx)^{n-r}(my)^{k-j}\,t^{k+mj}s^{n+mr}$$

(13)

By using Notation 1, the following generating matrix functions for Humbert matrix polynomials of two variables follows

$$=\sum_{n=0}^{\infty}\sum_{k=0}^{\infty}\sum_{r=0}^{\infty}\sum_{j=0}^{\infty}\frac{\left(A+(n+k)I\right)_{(r+j)}}{\left(A+(n+k)I\right)_{m(r+j)}\,r!\,j!\,n!\,k!}\times(-1)^{r+j}(mx)^{n-r}(my)^{k-j}\,t^{k+mj}s^{n+mr}$$

$$=\sum_{n=0}^{\infty}\sum_{k=0}^{\infty}\sum_{r=0}^{\infty}\sum_{j=0}^{\infty}\frac{\left(A+(n+k)I\right)_{(r+j)}}{r!\,j!\,n!\,k!\,m^{m}\prod_{s=1}^{m}\left(\frac{1}{m}\left(A_{A_{(n+k)}}+(s+1)I\right)\right)_{(r+j)}}\times(-1)^{r+j}(mx)^{n-r}(my)^{k-j}\,t^{k+mj}s^{n+mr}$$

$$\sum_{n=0}^{\infty}\sum_{k=0}^{\infty} P_{n,k,m}(x,y,A)\left[(A)_{n+k}\right]^{-1} s^n t^k$$

$$=\sum_{n=0}^{\infty}\sum_{k=0}^{\infty}\frac{(mxt)^n (mys)^k}{n!k!}\sum_{r=0}^{\infty}\sum_{j=0}^{\infty}\frac{\left(A+(n+k)I\right)_{(r+j)} t^{k+mj} s^{n+mr}}{\prod_{s=1}^{m}\left(\frac{1}{m}\left(A_{(n+k)}+(s+1)I\right)\right)_{(r+j)} r!j!}\times\left(\left(\frac{-s}{m}\right)^m\right)^r\left(\left(\frac{-t}{m}\right)^m\right)^j$$

have thus discovered the family of double generating function of the Humbert polynomials of two variables

$$\sum_{n=0}^{\infty}\sum_{k=0}^{\infty} P_{n,k,m}(x,y,A)\left[(A)_{n+k}\right]^{-1} s^n t^k=\sum_{n=0}^{\infty}\sum_{k=0}^{\infty}\frac{(mxt)^n (mys)^k}{n!k!}$$

$$\times {}_1F_m^{(2)}\left[\left(A+(n+k)I\right);\frac{A+(n+k)I}{m}I,\frac{A+((n+k)+1)I}{m}I,\cdots,\frac{A+((n+k)+(m-1))I}{m}I;\left(\frac{-s}{m}\right)^m,\left(\frac{-t}{m}\right)^m\right] \tag{14}$$

If B is a positive stable matrix in $C^{N\times N}$, then let us now return to (12) and consider the double sum.

$$\sum_{n=0}^{\infty}\sum_{k=0}^{\infty}\frac{(B)_{n+k} P_{n,k,m}(x,y,A) t^k s^n}{(A)_{n+k}}=\sum_{n=0}^{\infty}\sum_{k=0}^{\infty}\sum_{r=0}^{\left[\frac{n}{m}\right]}\sum_{j=0}^{\left[\frac{k}{m}\right]}\frac{(B)_{n+k}(A)_{(n+k)-(m-1)(r+j)}}{(A)_{n+k} r!j!(n-mr)!(k-mj)!}\times(-1)^{r+j}(mx)^{n-r}(my)^{k-j} t^k s^n$$

Then in similar manner, we get

$$\sum_{n=0}^{\infty}\sum_{k=0}^{\infty}(B)_{n+k}P_{n,k,m}(x,y,A)\left[(A)_{n+k}\right]^{-1}t^k s^n$$

$$=\sum_{n=0}^{\infty}\sum_{k=0}^{\infty}\frac{(mxt)^n(mys)^k}{n!k!}\times {}_mF_{m-1}^{(2)}\left[\frac{B+(n+k)I}{m};\frac{B+((n+k)+1)I}{m},\frac{B+((n+k)+(m-1))I}{m},\right.$$

$$\left.A+(n+k)I;\frac{A+(n+k)I}{m};\frac{A+((n+k)+1)I}{m},\frac{A+((n+k)+(m-1))I}{m};-s^m,-t^m\right]$$

(15)

5. EXPANSIONS OF $P_{n,k,m}(x,y,A)$ IN SERIES OF HERMITE $H_{n,k}(x,y,A)$

Now, we derive expansions of $P_{n,k,m}(x,y,A)$ in series of Hermite $H_{n,k}(x,y,A)$ According to [18], the expansion of $x^n I$ in a series of Hermite matrix polynomials was given in the form:

$$\frac{x^n}{n!}I=\left(\sqrt{2A}\right)^{-n}\sum_{k=0}^{[n/2]}\frac{y^k}{k!(n-2k)!}H_{n-2k}(x,y,A) \tag{16}$$

which with the aid of (5) and (9), one gets

$$\sum_{n=0}^{\infty}\sum_{k=0}^{\infty} P_{n,k,m}(x,y,A)t^k s^n = \sum_{n=0}^{\infty}\sum_{k=0}^{\infty}\sum_{r=0}^{\left[\frac{n}{m}\right]}\sum_{j=0}^{\left[\frac{k}{m}\right]} \frac{(A)_{(n+k)-(m-l)(r+j)}}{r!\,j!\,(n-mr)!\,(k-mj)!} \times (-1)^{r+j} (mx)^{n-mr} (my)^{k-mj} t^k s^n$$

$$= \sum_{n=0}^{\infty}\sum_{k=0}^{\infty}\sum_{r=0}^{\infty}\sum_{j=0}^{\infty} \frac{(-1)^{r+j} (A)_{(n+k)(r+j)}}{r!\,j!\,n!\,k!} \times (mx)^n (my)^k t^{k+mj} s^{n+mr}$$

From (16), we get

$$= \sum_{n=0}^{\infty}\sum_{k=0}^{\infty}\sum_{r=0}^{\infty}\sum_{j=0}^{\infty}\sum_{p=0}^{[n/2]} \frac{y^p \left(\sqrt{2A}\right)^{-n} (A)_{(n+k)(r+j)}}{r!\,j!\,k!\,p!\,(n-2p)!} \times (-1)^{r+j} (my)^k H_{n-2p}(mx,y,A) s^{n+mr} t^{k+mj}$$

$$= \sum_{n=0}^{\infty}\sum_{k=0}^{\infty}\sum_{r=0}^{\infty}\sum_{j=0}^{\infty}\sum_{p=0}^{\infty} \frac{y^p \left(\sqrt{2A}\right)^{-n-2p} (A)_{((n+2p)+k)+(r+j)}}{r!\,j!\,k!\,p!\,n!} \times (-1)^{r+j} (my)^k H_n(mx,y,A) s^{n+2p+mr} t^{k+mj}$$

$$= \sum_{n=0}^{\infty}\sum_{k=0}^{\infty}\sum_{r=0}^{\left[\frac{n}{m}\right]}\sum_{j=0}^{\left[\frac{k}{m}\right]}\sum_{p=0}^{\min(r,j)} \frac{(-1)^{(r-p)+(j-p)}\, y^p}{(r-p)!\,(j-p)!} \times \frac{\left(\sqrt{2A}\right)^{-n-2m-2p} (my)^{k-mj}}{(k-mj)!\,(n-mr)!\,p!} \times (A)_{(n+k)-(m-1)(r+j)} H_{n-mr}(mx,y,A) s^{n+(2-m)p} t^k$$

Now replacing n by $n+(m-2)p$ and equating the coefficients of $s^n \ t^k$, we get

$$P_{n,k,m}(x,y,A) = \sum_{r=0}^{[n+(m-2)p/m]}\sum_{j=0}^{\left[\frac{k}{m}\right]}\sum_{p=0}^{\min(r,j)} \frac{(-r)_p (-j)_p}{r!\,j!\,p!\,(k-mj)!} \times \frac{(-1)^{(r+j)} (m)^{k-mj} (y)^{k+p-mj}}{(n-2p-m(p+r))!}$$
$$\times \left(\sqrt{2A}\right)^{-n-m(p+r)} (A)_{(n+k)+(m-2p)-(m-l)(r+j)} \times H_{n-2p-m(r+p)}(mx,y,A) \tag{17}$$

REFERENCES

1. A. G. Constantine and R. J. Mairhead, "Partial Differential Equations for Hypergeometric Functions of Two Argument Matrices," Journal of Multivariate Analysis, Vol. 2, No. 3, 1972, pp. 332-338.

2. R. J. Muirhead, "Systems of Partial Differential Equations for Hypergeometric Functions of Matrix Argument," Annals of Mathematical Statistics, Vol. 40, No. 3, 1970, pp. 991-1001.

3. A. Terras, "Special Functions the Symmetric Space of Positive Matrices," SIAM Journal on Mathematical Analysis, Vol. 16, No. 3, 1985, pp. 620-640.

4. A. T. James, "Special Functions of Matrix and Single Argument in Statistics in Theory and Application of Special Functions," Academic Press, New York, 1975.

5. A. J. Duran, "Markov's Theorem for Orthogonal Matrix Polynomials," Canadian Journal of Mathematics, Vol. 48, 1996, pp. 1180-1195.

6. A. J. Duran and W. Van Assche, "Orthogonal Matrix Polynomials and Higher Order Recurrence Relations," Linear Algebra and Its Applications, Vol. 219, No. 1, 1995, pp. 261-280.

7. L. Jódar and J. Sastre, "The Growth of Laguerre Matrix Polynomials on Bounded Inervals," Applied Mathematics Letters, Vol. 13, No. 8, 2000, pp. 21- 26.

8. L. Jódar, R. Company and E. Navarro, "Laguerre Matrix Polynomials and Systems of Second Order Differential Equations," Applied Numerical Mathematics, Vol. 15, No. 1, 1994, pp. 53-63.

9. A. Sinap and W. Van Assch, "Orthogonal Matrix Polynomials and Applications," Journal of Computational and Applied Mathematics, Vol. 66, No. 1-2, 1996, pp. 27-52.

10. S. Saks and A. Zygmund, "Analytic Functions," Elsevier, Amsterdam, 1971.

11. N. Dunford and J. Schwartz, "Linear Operators, Part I," Interscience, New York, 1955.

12. L. Jódar and J. C. Cortés, "On the Hypergeometric Matrix Function," Journal of Computational and Applied Mathematics, Vol. 99, No. 1-2, 1998, pp. 205-217.

13. G. S. Khammash, "A Study of a Two Variables Gegenbauer Matrix Polynomials and Second Order Partial Matrix Differential Equations," International Journal of Mathematical Analysis, Vol. 2, No. 17, 2008, pp. 807-821.

14. E. Defez and L. Jódar, "Some Applications of the Hermite Matrix Polynomials Series Expansions," Journal of Computational and Applied Mathematics, Vol. 99, No. 1-2, 1998, pp. 105-117.

15. E. Hille, "Lectures on Ordinary Differential Equations," Addison-Wesley, New York, 1969.

16. L. Jódar and J. C. Cortés, "Some Properties of Gamma and Beta Matrix Functions," Applied Mathematics Letters, Vol. 11, No. 1, 1998, pp. 89-93.

17. E. D. Rainvlle, "Special Functions," Chelsea Publishing Co., Bronx, New York, 1960.

18. R. S. Batahan, "Anew Extension of Hermite Matrix Polynomials and Its Applications," Linear Algebra and Its Applications, Vol. 419, No. 1, 2006, pp. 82-92.

8

CHAPTER

Zero-M-Cordial Labeling of Some Graphs

Freeda Selvanayagom, Robinson S. Chellathurai

Department of Mathematics, Scott Christian College, Nagercoil, India

ABSTRACT

In this paper we prove that the complete bipartite graph $k_{m,n}$ where m and n are even, join of two cycle graphs c_n and c_m where $n+m\equiv 0$ (mod 4), split graph of c_n for even "n", $K_n \times P_2$ where n is even are admits a Zero-M-Cordial labeling. Further we prove that $B_n = K_{1,n} \times P_2$, $K_n \times P_2$ of odd n admits a Zero-M-Cordial labeling.

Keywords: Zero-M-Cordial Labeling; Split Graphs; Cartesian Product; H-Cordial

1. INTRODUCTION

We begin with finite, connected and undirected graph G = (V(G), E(G). If the vertices of the graph are assigned values subject to certain conditions then it is known as graph labeling. Any graph labeling will have the following three common characteristics. A set of numbers from which Vertex labels are chosen; $v_f(i)$ = number of vertices of G having label i under f

$e_f(i)$ = number of edges of G having label i under f*.

The concept of cordial labeling was introduced by I. Cahit, who called a graph G is Cordial if there is a vertex labeling $f: v(G) \rightarrow \{0,1\}$ such that the induced labeling

$f^*: E(G) \rightarrow \{0,1\}$, defined by

$f^*(xy) = |f(x) - f(y)|$, for all edges $xy \in E(G)$ and with the following inequalities holding and

$|v_f(0) - v_f(1)| \leq 1$ and $|e_f(0) - e_f(1)| \leq 1$.

In [1] introduced the concept of H-Cordial labeling. Cahit calls a graph H-Cordial if it is possible to label the edges with the numbers from the set $\{1,-1\}$ in such a way that, for some k, at each vertex v the sum of the labels on the edges incident with v is either k or -k and the inequalities $|v_f(k) - v_f(-k)| \leq 1$ and

$|e_f(1) - e_f(-1)| \leq 1$ are also satisfied where v(i)and e(j) are respectively, the number of vertices labeled with i and the number of edges labeled with j. He calls a graph H_n-Cordial if it is possible to label the edges with the numbers from the set $\{\pm 1, \pm 2, \cdots, \pm n\}$ in such a way that, at each vertex v the sum of the labels on the edges incident with v is in the set $\{\pm 1, \pm 2, \cdots, \pm n\}$ and the inequalities $|v_f(i) - v_f(-i)| \leq 1$ and $|e_f(i) - e_f(-i)| \leq 1$ are also satisfied for each i with $1 \leq i \leq n$. The concept of Zero-M-Cordial labeling is defined in [2]. A labeling f of a graph G is called Zero-M-Cordial, if for each vertex v, f(v) = 0. A graph G is called to be Zero-M-Cordial, if it admits a Zero-M-Cordial labeling. The usefulness of the above definition appears when one tries to find an HCordial labeling for a given graph G. If H is a ZeroM-Cordial subgraph of G then H-Cordiality of G\E(H) simply implies H-Cordiality G.

In [1] proved that $k_{n,n}$ is H-Cordial if and only if n > 2 and "n" is even; and $k_{m,n}$, m ≠ n is H-Cordial if and only if n ≡ 0 (mod4), m is even and m > 2, n > 2.

In [2] proved that k_n is H-Cordial if and only if n ≡ 0 or 3 (mod4) and n ≠ 3. W_n is H-Cordial if and only if n is odd. k_n is not H2-Cordial if n ≡ 1 (mod4). Also [2] prove that every wheel has an H_2-Cordial labeling.

In [3] several variations of graph labeling such as graceful, bigraceful, harmonious, cordial, equitable, humming etc. have been introduced by several authors. For definitions and terminologies in graph theory we refer to [4].

In this paper we investigate Zero-M-Cordial labeling on some Cartesian product of graphs, join of two graphs, and bipartite graph.

1.1. Definition: The join $G = G_1 + G_2$ of graph G_1 and G_2 with disjoint point sets V_1 and V_2 and edge sets E_1 and E_2 denoted by $G = G_1 + G_2$ is the graph union $G_1 \cup G_2$ together with all the edges joining v_1, v_2. If G_1 is (p_1,q_1) graph and G_2 is (p_2,q_2) graph then $G_1 + G_2$ is a $(p_1 + p_2, q_1 + q_2 + p_1 p_2)$.

1.2. Definition: Let $G_1 = (V_1, E_1)$ and $G_2 = (V_2, E_2)$ be two graphs. The Cartesian product of G_1 and G_2 which is denoted by $G_1 \times G_2$ is the graph with vertex set $v = v_1 \times v_2$ consisting of vertices $V = \{u = (u_1, u_2), v = (v_1, v_2) /$ u and v are adjacent in $G_1 \times G_2$ whenever $u_1 = v_1$ and u_2 adjacent to v_2 or u_1 adjacent to v_1 and $u_2 = v_2\}$.

1.3. Definition: For a graph G the split graph is obtained by adding to each vertex v, a new vertex v' such that v' is adjacent to every vertex that is adjacent to v in G. The resultant graph is denoted by spl (G).

2. MAIN RESULTS

Theorem 2.1: Every cycle C_n of even order admits a Zero-M-Cordial labeling Proof: Let $v_1, v_2, \cdots, v_n$ be the vertices of cycle C_n

Define f: $E(G) \to \{1, -1\}$ two cases are to be considered.

Case (i) n $\equiv$ 2 (mod 4)

For $1 \le i \le n$

$$f(v_i, v_{i+1}) = \begin{cases} = 1, \text{ if } i \equiv 1 \pmod{2} \\ = -1, \text{ if } i \equiv 0 \pmod{2} \text{ and } 1 \le i \le n \end{cases}$$

In view of the above labeling pattern we give the proof as follows:

When n $\equiv$ 2 (mod 4)

The total number of edges labeled with $-1's$ are given by $e_f(-1) = n/2$ and the total number of edges labeled with $1's$ are given by $e_f(1) = n/2$. Therefore the total difference between the edges labeled with $-1's$ and $1's$ are given by $|e_f(1) - e_f(-1)| = 0$. The induced vertex labels are equal to zero. Thus for each vertex v, $f(v) = 0$ and $|e_f(1) - e_f(-1)| \le 1$.

Case (ii) n $\equiv$ 0 (mod 4)

The total number of edges labeled with $-1's$ are given by $e_f(-1) = n/2$ and the total number of edges labeled with $1's$ are given by $e_f(1) = n/2$. Therefore the total difference between the edges labeled with $-1's$ and $1's$ are given by $|e_f(1) - e_f(-1)| = 0$. The induced vertex labels are equal to zero. Thus for each vertex v, $f(v) = 0$ and $|e_f(1) - e_f(-1)| \le 1$.

Hence the cycle graph cn, even n admits a zero-Mcordial labeling.

The vertex and the edge conditions are given in Table 1. The illustration is given in Figures 1 and 2.

In Figure 1 illustrates the Zero-M-Cordial labeling for the cycle graph C_6. Among the six edges three edges receive the label +1 and the other three edges receive the label −1. In Figure 2illustrates the Zero-M-Cordial labeling for the cycle graph C_8. Among the eight edges four edges receive the label +1 and the other four edges receive the label −1.

Theorem 2.2: The complete bipartite graph $k_{m,n}$ admits a Zero-M-Cordial labeling for all m, n such that m + n $\equiv$ 0, 2 (mod 4).

Proof: Let $v_1, v_2, \cdots, v_m$ and $u_1, u_2, \cdots, u_n$ are the vertex set of the bipartite graph $k_{m,n}$. The number of vertices and the edges of $k_{m,n}$ is m + n and mn respectively.

Define f: $E(G) \rightarrow \{1, -1\}$

Table 1. The vertex and the edge conditions of cycle graph C_n.

n	Vertex condition	Edge condition
$n \equiv 2 \pmod 4$	$f(v) = 0$	$e_f(1) = e_f(-1) = n/2$
$n \equiv 2 \pmod 4$	$f(v) = 0$	$e_f(1) = e_f(-1) = n/2$

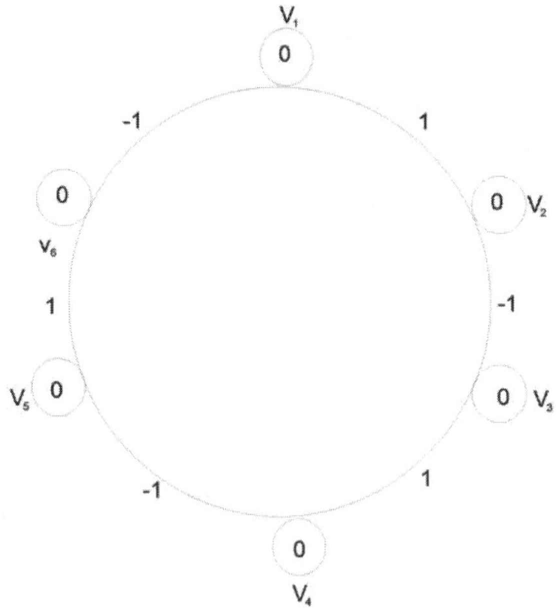

Figure 1. Zero-M-Cordial labeling on C_6.

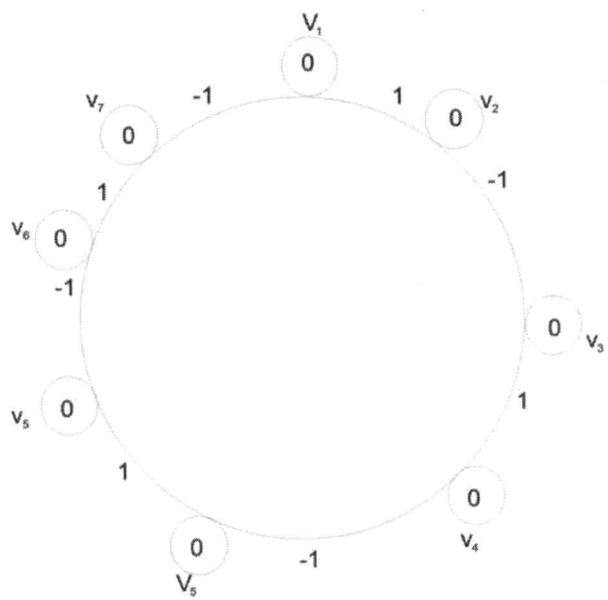

Figure 2. Zero-M-Cordial labeling on C_8.

The edge matrix of $k_{m,n}$ is given in Table 2.

In view of the above edge matrix we give the proof as follows.

Case (i) when $m + n \equiv 0 \pmod 4$, $m = n$.

Consider the bipartite graph $k_{4,4}$.

Using Table 2 the edge label matrix of $k_{4,4}$ is given by

$$
\begin{array}{cccccc}
 & u_1 & u_2 & u_3 & u_4 & \\
 & -1 & 1 & -1 & 1 & \\
0 & 1 & -1 & 1 & -1 & v_1 \\
0 & -1 & 1 & -1 & 1 & v_2 \\
0 & 1 & -1 & 1 & -1 & v_3 \\
0 & 0 & 0 & 0 & 0 & v_4
\end{array}
$$

Table 2. Edge matrix of $k_{m,n}$.

$$
\begin{array}{c|cccc|c}
 & u_1 & u_2 & \cdots & u_n & \\
v_1 & v_1u_1 & v_1u_2 & \cdots & v_1u_m & v_1 \\
v_2 & v_2u_1 & v_2u_2 & \cdots & v_2u_m & v_2 \\
\vdots & \vdots & \vdots & & \vdots & \vdots \\
\vdots & \vdots & \vdots & & \vdots & \vdots \\
v_m & v_mu_1 & v_mu_2 & \cdots & v_mu_n & v_m \\
 & u_1 & u_2 & \cdots & u_n &
\end{array}
$$

With respect to the above labeling the total number of edges labeled with $1'^s$ and $-1'^s$ are given by $e_f(1)=n/2$ and $e_f(-1)=n/2$. Therefore the total difference between the edges labeled with $-1'^s$ and $1'^s$ are given by $|e_f(1)-e_f(-1)|=0$. The induced vertex labels are equal to zero. Thus for each vertex v, $f(v)=0$ and $|e_f(1)-e_f(-1)|\le 1$.

Hence the bipartite graph $K_{4,4}$ admits a Zero-M-Cordial labeling. The vertex and the edge conditions are given in Table 3. The illustration is given in Figure 3.

Figure 3 illustrates the Zero-M-Cordial labeling on $K_{4,4}$. Among the Sixteen edges eight edges receive the label +1 and the other eight edges receive the label −1.

Case (ii) When m + n ≡ 2 (mod 4), $m \neq n$.

Consider the bipartite graph $K_{2,4}$.

Using Table 2 the edge label matrix is given by

$$\begin{array}{cccccc} & u_1 & u_2 & u_3 & u_4 & \\ & -1 & -1 & 1 & 1 & \\ 0 & 1 & 1 & -1 & -1 & v_1 \\ 0 & 0 & 0 & 0 & 0 & v_2 \end{array}$$

With respect to the above labeling the total number of edges labeled with $1'^s$ and $-1'^s$ are given by $e_f(1)=n/2, e_f(-1)=n/2$. Therefore the total difference between the edges labeled with $-1'^s$ and $1'^s$ are given by $|e_f(1)-e_f(-1)|=0$. The induced vertex labels are equal to zero. Thus for each vertex v, $f(v)=0$ and $|e_f(1)-e_f(-1)|\le 1$.

Hence the bipartite graph $k_{2,4}$ admits a zero-M-cordial labeling.

Figure 4 illustrates the Zero-M-Cordial labeling on $k_{2,4}$. Among eight edges four edges receive the label +1 and other four edges receive the label –1.

Case (iii) When m + n $\equiv$ 0 (mod 4) and $m \neq n$.

Consider the bipartite graph $k_{2,6}$.

Using Table 2 the edge label matrix is given by

$$\begin{array}{cccccccc} & u_1 & u_2 & u_3 & u_4 & u_5 & u_6 & \\ & 1 & -1 & 1 & -1 & 1 & -1 & \\ 0 & -1 & 1 & -1 & 1 & -1 & 1 & v_1 \\ 0 & 0 & 0 & 0 & 0 & 0 & 0 & v_2 \end{array}$$

With respect to the above labeling the total number of edges labeled with $1's$ and $-1's$ are given by

Table 3. The vertex and the edge conditions of the bipartite graph $K_{m,n}$.

n	Vertex condition	Edge condition
$m+n\equiv 0 \pmod 4, m=n$	$f(v)=0$	$e_f(1)=e_f(-1)=n/2$
$m+n\equiv 2 \pmod 4, m\neq n$	$f(v)=0$	$e_f(1)=e_f(-1)=n/2$
$m+n\equiv 0 \pmod 4$, and $m\neq n$	$f(v)=0$	$e_f(1)=e_f(-1)=n/2$

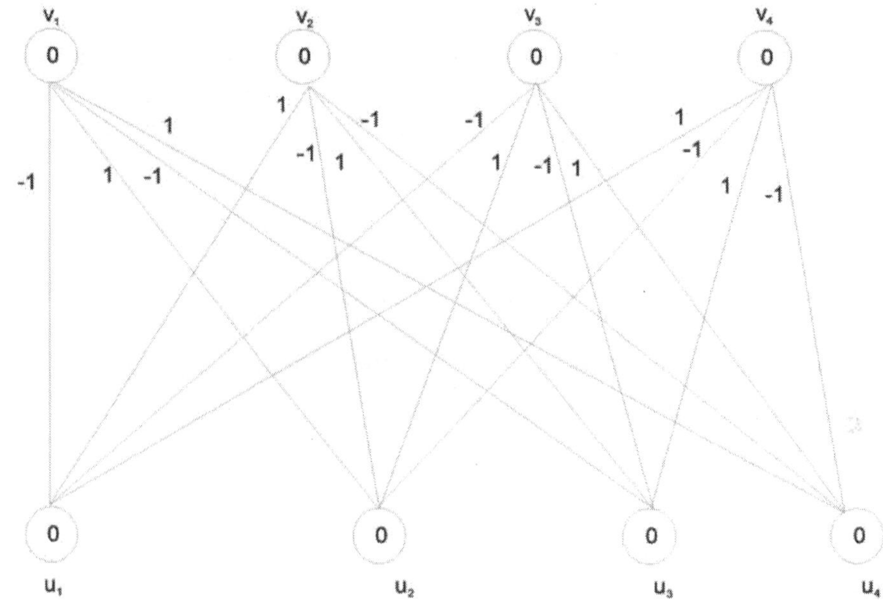

Figure 3. Zero-M-Cordial labeling on $k_{4,4}$.

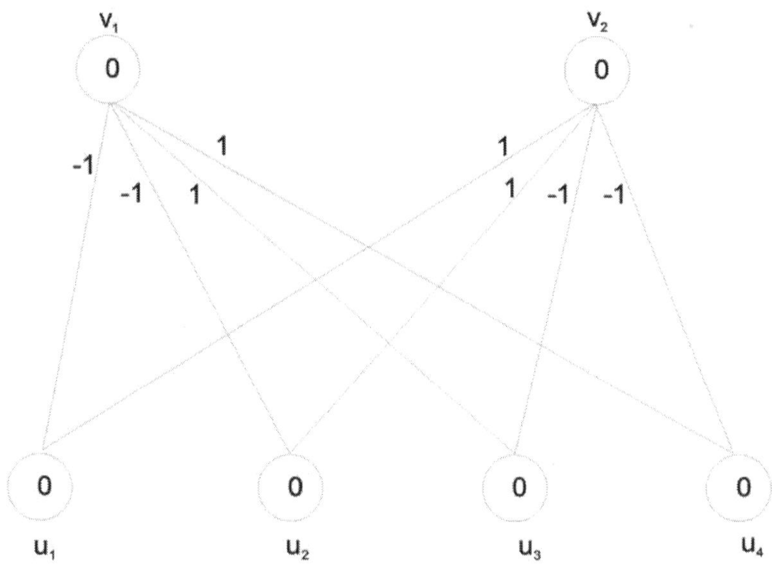

Figure 4. Zero-M-Cordial labeling on $k_{2,4}$.

$e_f(1) = n/2$ and $e_f(-1) = n/2$. Therefore the total difference between the edges labeled with $-1's$ and $1's$ are given by $|e_f(1) - e_f(-1)| = 0$. The induced vertex labels are equal to zero. Thus for each vertex v, $f(v) = 0$ and $|e_f(1) - e_f(-1)| \leq 1$.

Hence the bipartite graph $k_{2,6}$ admits a Zero-M-Cordial labeling. The vertex and the edge conditions are given in **Table 3**.

Figure 5 illustrates the Zero-M-Cordial labeling on $k_{2,6}$. Among the Twelve edges six edges receive the label +1 and the other six edges receive the label –1.

Theorem 2.3: The join of two cycle graphs C_n and C_m admits a Zero-M-Cordial labeling if n + m ≡ 0 (mod 4).

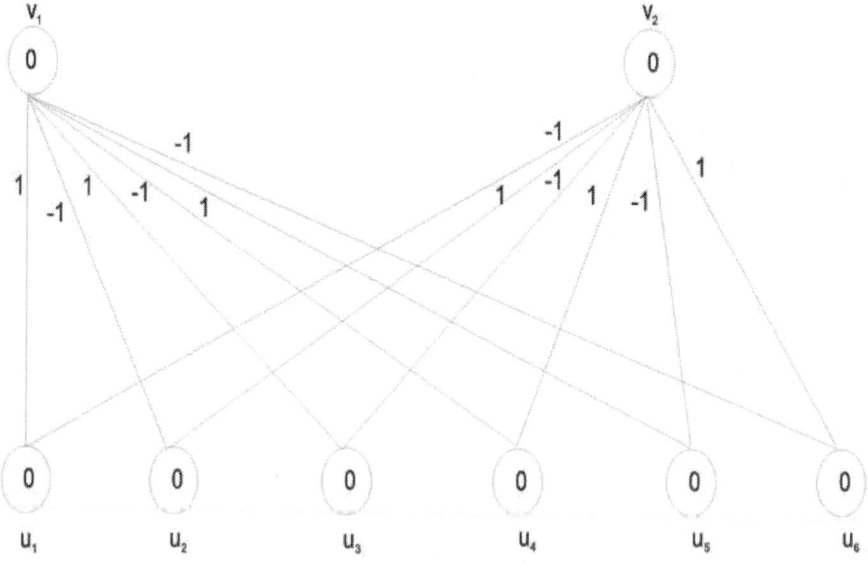

Figure 5. Zero-M-Cordial labeling on $k_{2,6}$.

Proof: Let $v_1, v_2, \cdots, v_n$ and $u_1, u_2, \cdots, u_m$ are the vertex set of the cycles C_n and C_m. The edge set E1 and E2 is the graph union of C_n and C_m together with all the edges joining the vertex set $v_1, v_2, \cdots, v_n$ and $u_1, u_2, \cdots, u_m$.

We note that $|V(G)| = p_1 + p_2$ and $|E(G)| = q_1 + q_2 + p_1 p_2$.

Define f: $E(G) \rightarrow \{1, -1\}$

The edge matrix of $c_n + c_m$ is given in **Table 4**.

In view of the above labeling pattern we give the proof as follows:

When n + m $\equiv$ 0 (mod 4)

Consider the join of two cycle graphs $C_4 + C_4$. Using **Table 4** the edge label matrix of C_4 & C_4 is given by

		u_1	u_2	u_3	u_4	
		-1	1	-1	1	
		-1	1	-1	1	
$1-1$	0	1	-1	1	-1	v_1
$1-1$	0	1	-1	1	-1	v_2
$-1\ 1$	0	$1-1$	$1-1$	-11	-11	v_3
$1\ -1$	0	0	0	0	0	v_4

With respect to the above labeling the total number of edges labeled with $1^{'s}$ and $-1^{'s}$ are given by $e_f(1) = n/2$ and $e_f(-1) = n/2$. Therefore the total difference between the edges labeled with $-1'^s$ and $1'^s$ are given by $|e_f(1)-e_f(-1)|=0$. The induced vertex labels are equal to zero. Thus for each vertex v, $f(v)=0$ and $|e_f(1)-e_f(-1)|\leq 1$.

Hence the join of two cycle graphs C_4 and C_4 admits a Zero-M-Cordial labeling. The vertex and the edge conditions are given in Table 5.

In Figure 6 illustrates the zero-M-Cordial labeling on $C_4 + C_4$. Among the twenty four edges twelve edges receive the label +1 and the other twelve edges label –1.

Theorem 2.4: The split graph of C_n, for even n, admits

Table 4. Edge matrix of $C_n + C_m$.

	u_1	u_2	$\cdots$	u_m	
v_1	v_1u_1	v_1u_2	$\cdots$	v_1u_m	$v_1v_2,\ v_1v_n$
v_2	v_2u_1	v_2u_2	$\cdots$	v_2u_m	$v_1v_2,\ v_2v_3$
$\vdots$	$\vdots$	$\vdots$		$\vdots$	$\vdots$
$\vdots$	$\vdots$	$\vdots$		$\vdots$	$\vdots$
v_n	v_nu_1	v_nu_2	$\cdots$	v_nu_m	$v_1v_n,\ v_{n-1}v_n$
	u_1u_2, u_1u_m	u_1u_2, u_2u_3	$\cdots$	$u_1u_m, u_{m-1}u_m$	

Table 5. The vertex and the edge conditions of two cycle graphs C_n and C_m.

C_n, C_m	Vertex condition	Edge condition
$m+n\equiv 0\ (\mathrm{mod}\ 4), m=n$	$f(v)=0$	$e_f(1)=e_f(-1)=n/2$

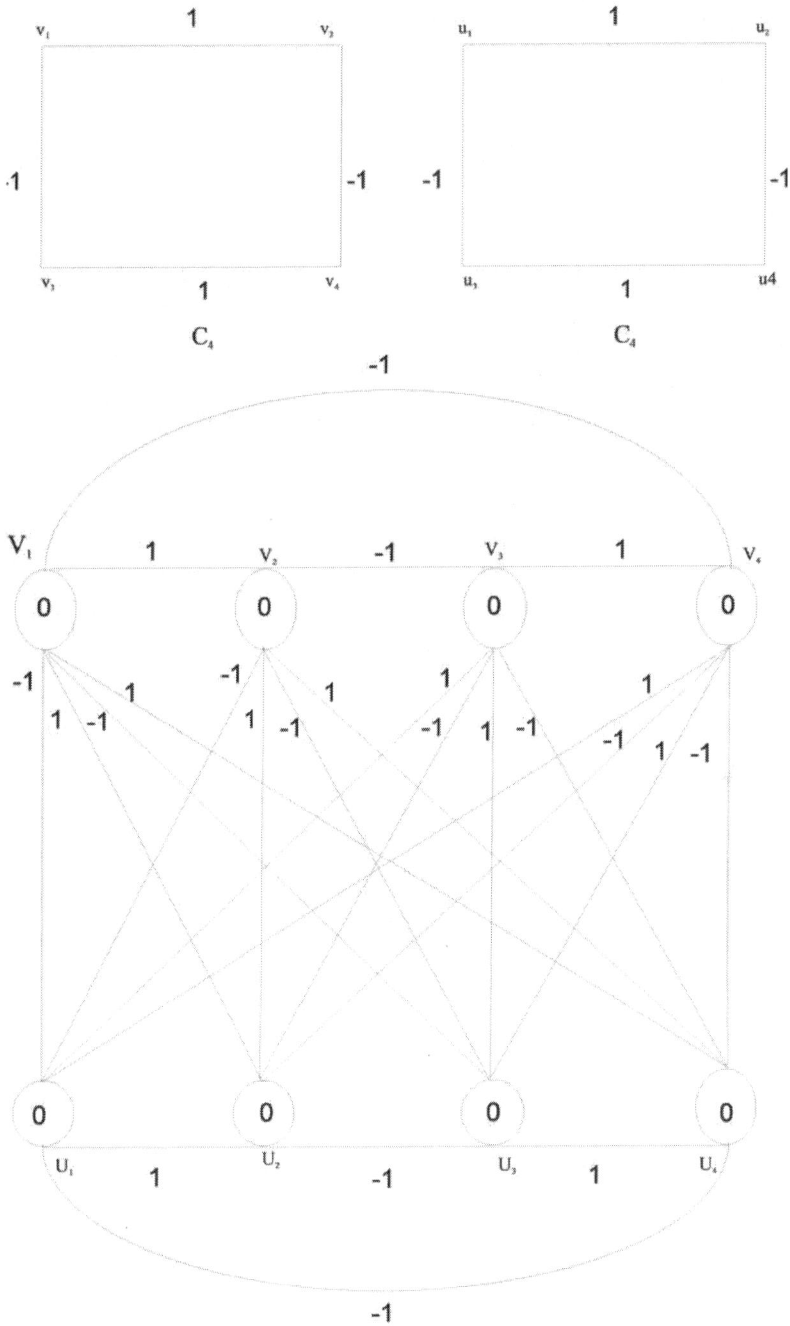

Figure 6. Zero-M-Cordial Labeling on $C_4 + C_4$.

a zero-M-cordial labeling.

Proof:

Let $v_1, v_2, \cdots, v_n$ be the vertices of cycle C_n and $v'_1, v'_2, \cdots, v'_n$ be the newly added vertices when n is even.

Let G be the split graph of cycle C_n with

$$V(G) = \{v_i, v'_i, 1 \le i \le n\},$$

$$E(G) = \{v_i, v'_{i+1}, 1 \le i \le n-1,\ v_n v_1, v'_i v_{i+1}, \\ v'_n v_1, v_i v'_{i+1}, 1 \le i \le n-1,\ v_n v'_1\}$$

we note that $|V(G)| = 2n$ and $|E(G)| = 3n$.

Define $f : E(G) \to \{1, -1\}$ two cases are to be considered.

Case (i) when n $\equiv$ 0 (mod 4)

For $1 \le i \le n-1$

$$f(v_i, v_{i+1}) = \begin{cases} 1, & \text{if } i \text{ is odd} \\ -1, & \text{if } i \text{ is even} \end{cases}$$

$$f(v_n, v_1) = -1$$

For $1 \le i \le n-1$

$$f(v_i, v'_{i+1}) = -1$$

$$f\left(v_n, v_1'\right) = -1$$

For $1 \le i \le n-1$

$$f\left(v_i', v_{i+1}\right) = 1$$

$$f\left(v_n', v_1\right) = 1$$

Case (ii) when n $\equiv$ 2 (mod 4)

For $1 \le i \le n-1$

$$f\left(v_i, v_{i+1}\right) = \begin{cases} 1\ , & \text{if } i \equiv 1 \pmod 2 \\ -1, & \text{if } i \equiv 0 \pmod 2 \end{cases}$$

$$f\left(v_n, v_1\right) = 1$$

For $1 \le i \le n-1$

$$f\left(v_i, v_{i+1}'\right) = -1$$

$$f\left(v_n, v_1'\right) = -1$$

For $1 \le i \le n-1$

$$f\left(v_i', v_{i+1}\right) = 1$$

$$f\left(v_n', v_1\right) = 1$$

With respect to the above labeling pattern we give the proof as follows.

The total number of edges labeled with $1^{'s}$ and $-1^{'s}$ are given by $e_f(1)=n$ and $e_f(-1)=n$. Therefore the total difference between the edges labeled with $-1^{'s}$ and $1^{'s}$ are given by $|e_f(1)-e_f(-1)|=|n-n|=0,$ differ by zero. The induced vertex labels are equal to zero. Thus for each vertex v, $f(v)=0$ and $|e_f(1)-e_f(-1)|\leq 1$.

Hence the split graph of C_n for even n admits a ZeroM-Cordial labeling. The vertex and the edge conditions are given in Table 6.

In Figure 7 illustrates the Zero-M-Cordial labeling on split c8. Among the twenty four edges twelve edges receive the label +1 and the other twelve edges receive the label –1.

Theorem 2.5: $K_n \times P_2$ admits a Zero-M-Cordial labeling for even n.

Proof: Let G be the graph $K_n \times P_2$ where n is even and $V(G)=\{V_{ij} \quad i=1,2,\cdots,n \text{ and } j=1,2\}$ be the vertices of the graph G.

We note that $|V(G)|=2n$ and $|E(G)|=n^2$ as

$|V(k_n)|=n$ and $|E(k_n)|=\dfrac{n(n-1)}{2}$

Define $f:E(G)\rightarrow\{1,-1\}$ as follows For $1\leq i,k\leq n$

$$f\left(v_{i_1}, v_{k_1}\right) = 1$$

For $n < i,k \leq n+2$

$$f\left(v_{i_1}, v_{k_1}\right) = -1$$

For $1 \leq i,k \leq n$

$$f\left(v_{i_2}, v_{k_2}\right) = 1$$

For $n < i,k \leq n+2$

$$f\left(v_{i_2}, v_{k_2}\right) = -1$$

For $1 \leq i \leq n$

$$f\left(v_{i_1}, v_{i_2}\right) = -1$$

With respect to the above labeling pattern we give the proof as follows.

The total number of edges labeled with $1'^s$ and $-1'^s$ are given by $e_f(1) = n/2$ and $e_f(-1) = n/2$. Therefore the total difference between the edges labeled with $-1'^s$ and $1'^s$ are given by $\left|e_f(1) - e_f(-1)\right| = 0,$ differ by 0. The induced vertex labels are equal to zero. Thus for each vertex v, $f(v) = 0$ and $\left|e_f(1) - e_f(-1)\right| \leq 1$.

Hence $K_n \times P_2$ admits a Zero-M-Cordial labeling for even n. The vertex and the edge conditions are given in Table 7.

Figure 8 illustrates the Zero-M-Cordial labeling on $k_n \times p_2$. Among the sixteen edges eight edges label +1 and the other eight edges label –1.

Theorem 2.6: $W_n \times P_2$ admits a Zero-M-Cordial labeling for odd n.

Proof: Let G be the graph $W_n \times P_2$ where n is odd and $V(G)=\{V_{ij} \quad i=1,2,\cdots,n+1 \text{ and } j=1,2\}$ be the vertices of graph G.

We note that $|V(G)|=2(n+1)$ and $|E(G)|=5n+1$

as $|V(W_n)|=n+1$ and $|E(W_n)|=2n$

Table 6. The vertex and the edge conditions of split of C_n.

C_n, C_m	Vertex condition	Edge condition
$n \equiv 0 \pmod 4$	$f(v)=0$	$e_f(1)=e_f(-1)=n$
$n \equiv 2 \pmod 4$	$f(v)=0$	$e_f(1)=e_f(-1)=n$

Table 7. Vertex and the edge condition of $k_n \times p_2$.

n	Vertex condition	Edge condition
Even	$f(v)=0$	$e_f(1)=e_f(-1)=n/2$

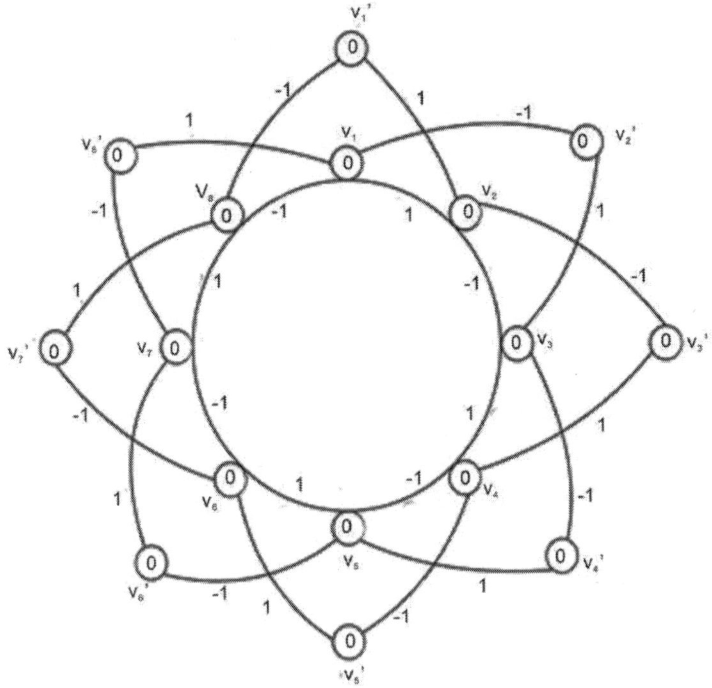

Figure 7. Zero-M-Cordial labeling on split C_8.

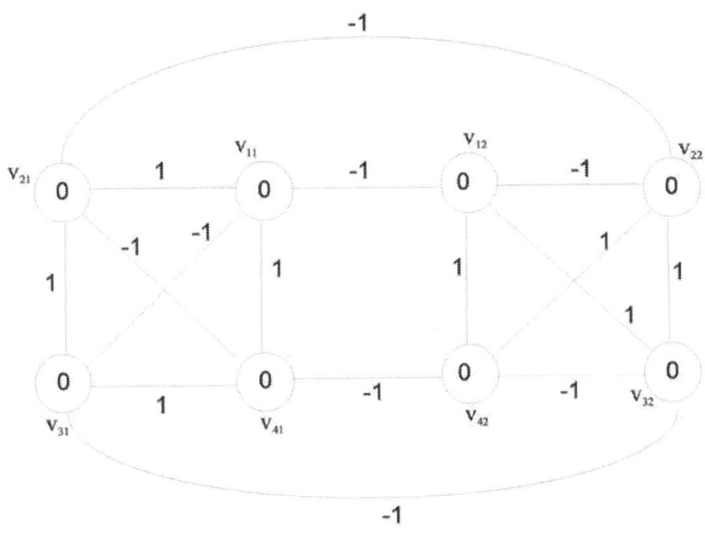

Figure 8. Zero-M-Cordial labeling on $k_4 \times p_2$.

Define $f:E(G)\rightarrow\{1,-1\}$ as follows For $1\leq i,k\leq 2n-2$

$$f\left(v_{i_1},v_{k_1}\right)=1$$

For $2n-2<i,k\leq 2n$

$$f\left(v_{i_1},v_{k_1}\right)=-1$$

For $1\leq i,k\leq 2n-2$

$$f\left(v_{i_2},v_{k_2}\right)=1$$

For $2n-2<i,k\leq 2n$

$$f\left(v_{i_2},v_{k_2}\right)=-1$$

For $1\leq i\leq n+1$

$$f\left(v_{i_1},v_{i_2}\right)=-1$$

The total number of edges labeled with $1'^s$ and $-1'^s$ are given by $e_f(1)=n/2$ and $e_f(-1)=n/2$. Therefore the total difference between the edges labeled with $-1'^s$ and $1'^s$ are given by $|e_f(1)-e_f(-1)|=0,$ differ by 0. The induced vertex labels are equal to zero. Thus for each vertex v, $f(v)=0$ and $|e_f(1)-e_f(-1)|\leq 1$.

Hence $W_n \times P_2$ admits a Zero-M-Cordial labeling for odd n. The vertex and the edge conditions are given in **Table 8**.

Figure 9 illustrates the Zero-M-Cordial labeling on $W_3 \times P_2$. Among the sixteen edges eight edges receive the label +1 and the other eight edges label –1.

Theorem 2.7: $B_n = k_{1,n} \times P_2$ (also known as book graph) admits a Zero-M-Cordial labeling for odd n.

Proof: Let G be the graph $K_{1,n} \times P_2$ where n is odd and $V(G) = \{V_{ij} \quad i = 1, 2, \cdots, n+1, j = 1, 2\}$ be the vertices of G.

We note that $|V(G)| = 2(n+1)$ and $|E(G)| = 3n+1$.

Define $f : E(G) \to \{1, -1\}$ as follows For $1 \le i, k \le n-1$

$$f\left(v_{i_1}, v_{k_1}\right) = 1$$

For $n-1 < i, k \le n$

$$f\left(v_{i_1}, v_{k_1}\right) = -1$$

For $1 \le i, k \le n-1$

$$f\left(v_{i_2}, v_{k_2}\right) = 1$$

For $n-1 < i, k \le n$

$$f\left(v_{i_2}, v_{k_2}\right) = -1$$

For $1 \leq i \leq n$

$$f\left(v_{i_1}, v_{i_2}\right) = -1$$

For $n < i \leq n+1$

$$f\left(v_{i_1}, v_{i_2}\right) = 1$$

The total number of edges labeled with $1'^s$ and $-1'^s$ are given by $e_f(1) = n/2$ and $e_f(-1) = n/2$. Therefore the total difference between the edges labeled with $-1'^s$ and $1'^s$ are given by $\left|e_f(1) - e_f(-1)\right| = \left|\frac{n}{2} - \frac{n}{2}\right| = 0,$

differ by 0. The induced vertex labels are equal to zero. Thus for each vertex v, $f(v) = 0$ and

Table 8. Vertex and the edge condition of $W_n \times P_2$.

n	Vertex condition	Edge condition
Odd	$f(v) = 0$	$e_f(1) = e_f(-1) = n/2$

Table 9. Vertex and the edge condition of $B_n = k_{1,n} \times P_2$.

n	Vertex condition	Edge condition
Odd	$f(v)=0$	$e_f(1)=e_f(-1)=n/2$

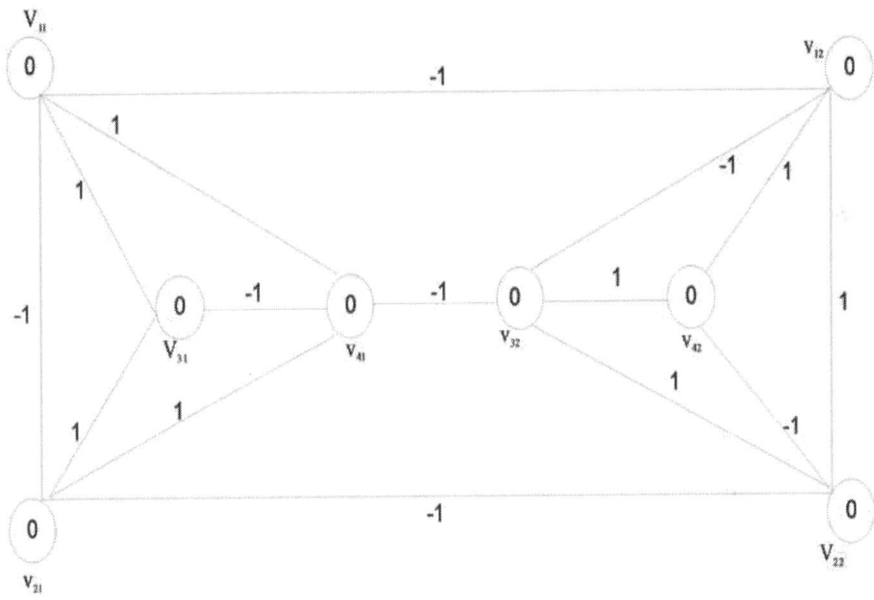

Figure 9. Zero-M-Cordial labeling on $W_3 \times P_2$.

$$\left|e_f(1)-e_f(-1)\right| \le 1.$$

Hence $B_n = k_{1,n} \times P_2$ (also known as book graph) admits a Zero-M-Cordial labeling for odd n. The vertex and the edge conditions are given in Table 9.

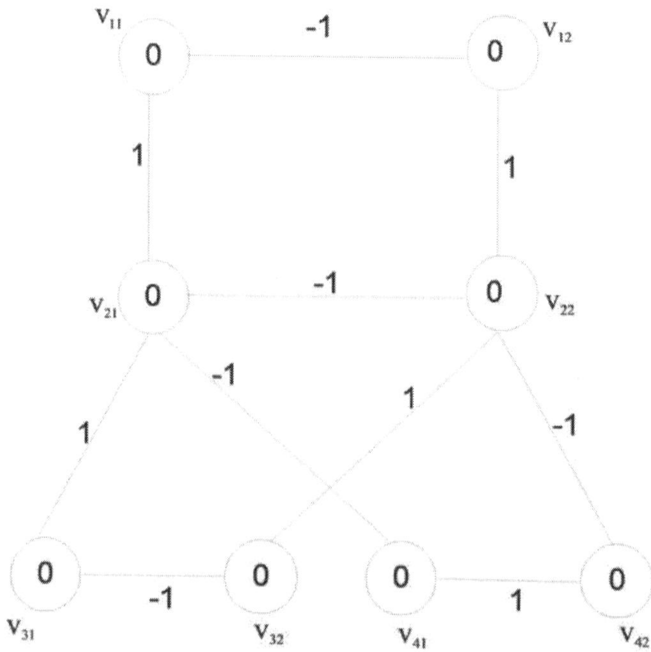

Figure 10. Zero-M-Cordial labeling on $K_{1,3} \times P_2$.

Figure 10 illustrates Zero-M-Cordial labeling on $K_{1,3} \times P_2$. Among the ten edges five edges receive the label +1 and the other five edges receive the label −1.

3. CONCLUDING REMARK

Here we investigate Zero-M-Cordial labeling for Cartesian product of some graphs, join of two cycle graphs, split graphs and bipartite graphs. Similar results can be derived for other graph families and in the context of different graph labeling problem is an open area of research.

REFERENCES

1. I. Cahit, "H-Cordial Graphs," Bulletin of the Institute of Combinatorics and Its Applications, Vol. 18, 1996, pp. 87-101.

2. M. Ghebleh and R. Khoeilar, "A Note on 'H-Cordial graphs'," Bulletin of the Institute of Combinatorics and Its Applications, Vol. 31, 2001, pp. 60-68.

3. J. A. Gallian, "A Dynamic Survey of Graph Labeling," The Electronics Journal of Combinatories, Vol. 18, 2011.

4. F. Harary, "Graph Theory," Addison Wesley, Reading, 1972.

9
CHAPTER

On q-Deformed Calculus in Quantum Geometry

Olaniyi S. Maliki[1], Emmanuel I. Ugwu[2]

[1]Department of Industrial Mathematics and Applied Statistics, Ebonyi State University, Abakaliki, Nigeria

[2]Department of Industrial Physics, Ebonyi State University, Abakaliki, Nigeria

ABSTRACT

The relation between noncommutative (or quantum) geometry and the mathematics of spaces is in many ways similar to the relation between quantum physics and classical physics. One moves from the commutative algebra of functions on a space (or a commutative algebra of classical observable in classical physics) to a noncommutative algebra representing a noncommutative space (or a noncommutative algebra of quantum observables in quantum physics). The object of this paper is to study the basic rules governing q-calculus as compared with the classical NewtonLeibnitz calculus.

Keywords: Quantum Geometry, q-Numbers, q-Factorials, q-Calculus

1. INTRODUCTION

There exists an intimate relationship between classical geometry and physics, and to appreciate this we will consider the Einstein field equations (EFE) of special relativity written as:

$$G_{\mu\nu} = 8\pi T_{\mu\nu} \tag{EFE}$$

where $G_{\mu\nu} = R_{\mu\nu} - \frac{1}{2} g_{\mu\nu} R$ is the Einstein tensor, $R = g^{\mu\nu} R_{\mu\nu}$ is the Ricci tensor. $T_{\mu\nu}$ is the energy-momentum tensor.

For the moment we are only interested in three basic properties of the above equation.

1) The equation (EFE) is a tensor equation. This is necessarily so, since the principle of invariance under coordinate transformations must hold, in other words the equations of physics must look the same in any frame of reference.

2) We can interpret equation (EFE) more simply as

Tensor representing geometry of space = Tensor representing energy content of space

i.e. it is the presence of matter in space that distorts the neighbouring geometry. Most equations of mathematical physics can be interpreted similarly.

3) The solution to equation (EFE) is a geometrical object, namely a line element given by

$$ds^2 = g_{\mu\nu} dx^\mu dx^\nu$$

where $g_{\mu\nu}$ is the metric tensor to be solved for in (EFE).

1.1. Quantum Geometry

Every geometry is associated with some kind of space. Quantum (or noncommutative) geometry [1][2] deals with quantum spaces, including the classical concept of space as a very special case. In classical geometry spaces are always regarded as collections of points equipped with the appropriate additional structure (as for example a topological structure given by the collection of open sets, or a smooth structure given by the atlas). In contrast to classical geometry, quantum spaces are not interpretable in this way. In general, quantum spaces have no points at all! They exhibit non-trivial quantum fluctuations' of geometry at all scales.

In generalizing classical geometry to the non-commutative level, there are two important conceptual steps:

1) Translation of geometry into a commutative algebra format;

2) Non-commutative generalizations.

1.2. Reformulating Basic Geometrical Concepts

It turns out that the geometrical structure on any given topological space X is always completely expressible in the language of some associated *-algebra [3] [4] .

Points

Let X be a compact topological space, and let $A = C(X)$ be the *-algebra of continuous complex-valued functions on X. Every element $x \in X$ naturally gives rise to a linear functional $\chi = \chi_x\colon A \to \mathbb{C}$ defined by

$$\chi(f) = f(x)$$

This map is multiplicative in the sense that

$$\chi(fg) = \chi(f)\chi(g), \qquad \forall f, g \in A$$

Furthermore, χ is Hermitian in the sense that $\chi(f^*) = \chi(f)^*$, it is also non-zero, i.e. χ is a character on A. Conversely, consider an arbitrary character χ, then it can be shown that there exists a unique point $x \in X$ such that $\chi = \chi_x$. In other words, we have a natural bijection between points of X and characters of A. It is important to note that this characterization of points also remains valid at the smooth level, in which X could be a compact smooth manifold and the associated *-algebra consists of smooth functions on X.

1.3. The Gelfand-Naimark Theorem

The algebra $A = C(X)$ of complex-valued functions on a compact topological space X, equipped with the maximum norm

$$|f| = \max_{x \in X} |f(x)|$$

is a commutative C*-algebra [2] . The classical theorem of Gelfand and Naimark characterizes the algebras of the form $A = C(X)$, as commutative unital C*-algebras. This means that for every commutative unital C^*-algebra A there exists (up to homeomerphisms) a unique compact topological space X such that $A \cong C(X)$.

As we have earlier observed, the points of the space X are recovered as characters of the associated algebra A. In terms of this identification, the topology on X coincides with the weak*-topology, induced from the dual space A^*, consisting of continuous linear functionals on A. It turns out that homomorphisms between C^*-algebras are automatically continuous, in particular characters are continuous linear functionals.

2. THE QUANTUM PLANE

A simple example of quantum geometry is the quantum plane (Figure 1). Usually, a plane is described by two coordinate functions x, y. Naturally, the functions xy and yx are the same since it does not matter whether you measure x first and then y or y first and then x. This is precisely what is lost in the quantum world.

In the quantum plane we replace the property $xy = yx$ by $xy = qyx$ where q is some parameter. We no longer have points, however we can continue to work algebraically with x and y.

Define $[x, y] = xy - yx$, the commutator bracket. Hence $xy = qyx$ can be rewritten as

$$[x, y] = (q - 1)\, yx$$

The commutative case is obtained when $q = 1$.

2.1. q-Deformed Calculus

It is interesting to know that one can really do geometry in this setting, where coordinates do not commute. This is the remarkable discovery in recent times. For example, we can follow the approach of Newton-Leibnitz defining differentiation by

$$f'(y) = x^{-1}\left(f(x+y) - f(y)\right)\Big|_{x\to 0}$$

But when $xy \neq yx$ and in particular $xy = qyx$ we get instead

$$[f'(x)]_q = \frac{f(x) - f(qx)}{(1-q)x}$$

Thus for example the derivative of the function $f(x) = x^n$ in this non-commutative setting would be

$$[f'(x)]_q = \frac{x^n - q^n x^n}{(1-q)x} = \left(\frac{1-q^n}{1-q}\right)x^{n-1} = \left(1+q+q^2+\cdots+q^{n-1}\right)x^{n-1}$$

We observe here that when $q = 1$ the derivative of x^n for the commutative case is obtained, i.e. $f'(x) = nx^{n-1}$.

2.2. Basic Notions of q-Calculus

The mathematical study of noncommutative geometry is intimately related to the so-called q-calculus (qnumbers, q-factorials, q-differentials and integrals, basic q-hypergeometric functions, and q-orthogonal polynomials). Here we give a brief introduction to q-numbers and q-factorials which will be required in the subsequent sections.

2.2.1. q-Numbers and q-Factorials

For any nonzero complex number q, the q-number $[a]_q$, $a \in \square$, is defined by

$$[a]_q \equiv [a] := \frac{q^a - q^{-a}}{q - q^{-1}} = \frac{e^{ah} - e^{-ah}}{e^h - e^{-h}} = \frac{\sinh ah}{\sinh h}, \qquad q = \exp h \tag{2.1}$$

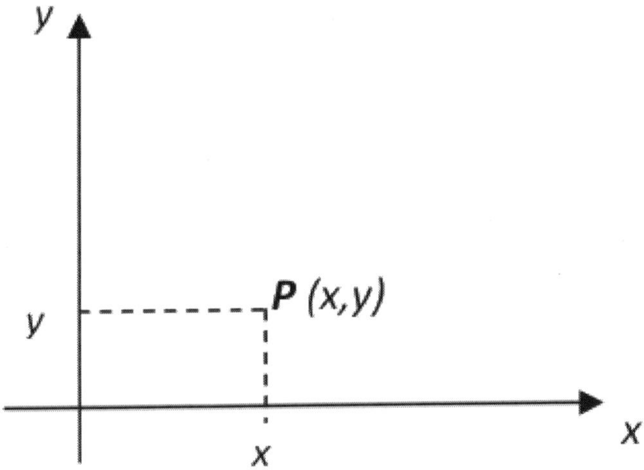

Figure 1. The quantum plane (xy ≠ yx).

We observe that, $\lim_{q \to 1} [a]_q = a$. The following expression which is easily proved will prove useful.

$$\Box a\Box_q \equiv \Box a\Box := \frac{1-q^a}{1-q} = 1+q+q^2+\cdots+q^{a-1} = q^{(a-1)/2}\left[a\right]_{q^{1/2}}$$

(2.2)

Thus given $f(x)=x^n$, as shown previously $\left[f'(x)\right]_q = \Box n\Box x^{n-1}$.

2.2.2. Proposition

The q-numbers satisfy the following relations derived from the property of the exponential function

1) $[r]=q^{r-1}+q^{r-3}+\cdots+q^{-(r-1)}$

2) $[r+s]=q^s[r]+q^{-r}[s]=q^{-s}[r]+q^r[s]$

3) $[r-s]=q^s[r]-q^r[s]=q^{-s}[r]-q^{-r}[s]$

4) $[r][s-t]+[s][t-r]+[t][r-s]\equiv 0$

5) $[r]=[2][r-1]-[r-2]$

The proof of 1) is easy to see from the fact that;

$$[r]=\frac{q^r-q^{-r}}{q-q^{-1}}=\frac{q^r\left(1-q^{-2r}\right)}{q\left(1-q^{-2}\right)}=q^{r-1}\left(\frac{1-\left(q^{-2}\right)^r}{1-\left(q^{-2}\right)}\right)$$

Let $u = q^{-2}, \ \therefore \ \dfrac{1-\left(q^{-2}\right)^r}{1-\left(q^{-2}\right)} = \dfrac{1-u^r}{1-u} = 1+u+u^2+\cdots+u^{r-1}$

Hence;

$$[r] = q^{r-1}\left(1+q^{-2}+q^{-4}+\cdots+q^{-2(r-1)}\right) = q^{r-1}+q^{r-3}+\cdots+q^{-(r-1)} \quad \square$$

The rest of the identities can be proved similarly. It is important to note that the relations 1)-5) remain valid when q is considered an indeterminate. Consequently any q-number $[n]_q, n \in \square$, belongs to the space

$\square\left[q, q^{-1}\right]$ of Laurent polynomials [2] in q with integral coefficients.

Suppose $n \in \square$, then we define the q-factorial $[n]_q! \equiv [n]!$ by setting;

$$[n]! = [1][2]\ldots[n], \qquad [0]! := 1$$

The following expression is quite useful in the theory of hypergeometric functions as well as in combinatorics.

$$\{a:q\}_n := \prod_{j=1}^{n}\left(1-aq^{j-1}\right) = (1-a)(1-aq)\left(1-aq^2\right)\cdots\left(1-aq^{n-1}\right), \quad \{a:q\}_0 = 1. \tag{2.3}$$

It is now possible for us to relate the above with the q-factorials. We observe that;

$$[n]! = \frac{q^{-n(n-1)/2}}{\left(1-q^2\right)^n}\left\{q^2 : q^2\right\}_n$$

(2.4)

From equation (2.3) we note that $\left\{q^{-m} : q\right\}_n = 0,\ \forall n = m+1, m+2, \cdots$ For $|q| < 1$, define

$$\{a:q\}_\infty := \lim_{n\to\infty}\{a:q\}_n = \prod_{j=1}^{\infty}\left(1-aq^{j-1}\right)$$

(2.5)

which converges $\forall a \in$ □ , and defines an analytic function on □ .

2.2.3. Proposition

$\forall a \in$ □ , $|q| < 1$, $\{a:q\}_n = \dfrac{\{a:q\}_\infty}{\{aq^n:q\}_\infty}$.

Proof

$$\frac{\{a:q\}_\infty}{\{aq^n:q\}_\infty} = \frac{\prod_{j=1}^{\infty}\left(1-aq^{j-1}\right)}{\prod_{j=1}^{\infty}\left(1-aq^{n+j-1}\right)} = \frac{(1-a)(1-aq)\left(1-aq^2\right)\left(1-aq^3\right)\cdots\left(1-aq^{n-1}\right)\left(1-aq^n\right)\left(1-aq^{n+1}\right)\cdots}{\left(1-aq^n\right)\left(1-aq^{n+1}\right)\left(1-aq^{n+2}\right)\cdots\left(1-aq^{2n-1}\right)\cdots}$$

$$= (1-a)(1-aq)\left(1-aq^2\right)\left(1-aq^3\right)\cdots\left(1-aq^{n-1}\right) = \{a:q\}_n$$

Remark: We can also define and show that;

$$\{a:q\}_{-n} = \frac{\{a:q\}_\infty}{\left\{aq^{-n}:q\right\}_\infty} = \frac{1}{\left\{aq^{-n}:q\right\}_n}$$

(2.6)

2.2.4. q-Binomial Coefficients

The q-binomial coefficients are defined by the formula;

$$\begin{Bmatrix} n \\ m \end{Bmatrix}_q \equiv \begin{Bmatrix} n \\ m \end{Bmatrix} := \frac{\{q:q\}_n}{\{q:q\}_m \{q:q\}_{n-m}} = \frac{[n]_{q^{1/2}}!\, q^{(n-m)m/2}}{[m]_{q^{1/2}}!\, [n-m]_{q^{1/2}}!}, \quad \forall n,m \in \mathbb{N} \cup \{0\},\ n \geq m. \tag{2.7}$$

Remark: There exist a close analogy between the classical binomial coefficients $\binom{n}{m} = \frac{n!}{(n-m)!m!}$ and their q-analogues. Many of the identities satisfied by the former have their counterparts for the q-binomial coefficients. For example the classical identity;

$$\binom{n+1}{m} = \binom{n}{m-1} + \binom{n}{m}$$

simply translates to;

$$\begin{Bmatrix} n+1 \\ m \end{Bmatrix}_q = \begin{Bmatrix} n \\ m-1 \end{Bmatrix}_q + q^m \begin{Bmatrix} n \\ m \end{Bmatrix}_q = q^{n-m+1} \begin{Bmatrix} n \\ m-1 \end{Bmatrix}_q + \begin{Bmatrix} n \\ m \end{Bmatrix}_q \tag{2.8}$$

2.2.5. Proposition

Let x and y be noncommuting variables satisfying the relation $xy = qyx$, then we have

$$(x+y)^n = \sum_{m=0}^{n} \begin{Bmatrix} n \\ m \end{Bmatrix}_q y^m x^{n-m} = \sum_{m=0}^{n} \begin{Bmatrix} n \\ m \end{Bmatrix}_{q^{-1}} x^m y^{n-m} \qquad (2.9)$$

In case q is a primitive pth root of unity, and p is odd, then

$$(x+y)^p = x^p + y^p \qquad (2.10)$$

Proof

Equation (2.9) can be established by induction on n, and employing the first identity in (2.8). The second assertion follows directly from (2.9). Observe that:

$$(x+y)^p = \sum_{m=0}^{p} \begin{Bmatrix} p \\ m \end{Bmatrix}_q y^m x^{p-m} = \begin{Bmatrix} p \\ 0 \end{Bmatrix}_q x^p + \begin{Bmatrix} p \\ p \end{Bmatrix}_q y^p + \sum_{m=1}^{p-1} \begin{Bmatrix} p \\ m \end{Bmatrix}_q y^m x^{p-m}$$

From (2.7)

$$\begin{Bmatrix} p \\ m \end{Bmatrix}_q = \frac{\{q:q\}_p}{\{q:q\}_m \{q:q\}_{p-m}} = \frac{[p]_{q^{1/2}}! q^{(p-m)m/2}}{[m]_{q^{1/2}}! [p-m]_{q^{1/2}}!} = 0, \quad m = 1,2,\dots,p-1.$$

and $\begin{Bmatrix} p \\ 0 \end{Bmatrix}_q = \begin{Bmatrix} p \\ p \end{Bmatrix}_q = 1$.

3. Q-DIFFERENTIAL AND Q-INTEGRAL OPERATORS

The following are important basic notions derived from their analogue in classical calculus, and will be employed subsequently.

3.1. The q-Differential Operator

For $1 \neq q \in \square$, we define the q-differential operator D_q by:

$$D_q f(x) = \frac{f(x) - f(qx)}{(1-q)x} \qquad (3.1)$$

Note that $D_q \to d/dx$ as $q \to 1$.

Proposition:

$$D_q f(x) = \sum_{r=0}^{\infty} \frac{(q-1)^r}{(r+1)!} x^r \frac{d^{r+1}}{dx^{r+1}} f(x), \quad \forall 1 \neq q \in \square \qquad (3.2)$$

Provided the expression on the right hand side exists.

Proof

Let $x + h = qx$, then by Taylor's series,

$$f(x+h) = f(x) + hf'(x) + \frac{h}{2!} f''(x) + \cdots$$

$$\therefore \quad f(x+h) - f(x) = hf'(x) + \frac{h}{2!} f''(x) + \cdots = \sum_{r=0}^{\infty} \frac{h^{r+1}}{(r+1)!} \frac{d^{r+1}}{dx^{r+1}} f(x)$$

Setting $h = (q-1)x$, we have:

$$D_q f(x) = \frac{f(qx) - f(x)}{(q-1)x} = \frac{1}{(q-1)x} \sum_{r=0}^{\infty} \frac{\left[(q-1)x\right]^{r+1}}{(r+1)!} \frac{d^{r+1}}{dx^{r+1}} f(x) = \sum_{r=0}^{\infty} \frac{(q-1)^r x^r}{(r+1)!} \frac{d^{r+1}}{dx^{r+1}} f(x)$$

The formula for the product of two functions is given by

$$D_q\{u(x)v(x)\}=\frac{u(x)v(x)-u(qx)v(qx)}{x(1-q)}=\frac{v(x)\{u(x)-u(qx)\}+u(qx)\{v(x)-v(qx)\}}{x(1-q)}$$

We can now define the q-analogue of the Newton-Leibnitz rule:

$$D_q\{u(x)v(x)\}=v(x)d_qu(x)+u(qx)d_qv(x) \tag{3.3}$$

Thus,

$$D_q^2f(x)=D_q(D_qf(x))=D_q\left(\sum_{r=0}^{\infty}\frac{(q-1)^r x^r}{(r+1)!}\frac{d^{r+1}}{dx^{r+1}}f(x)\right)$$

$$=\sum_{r=0}^{\infty}\frac{(q-1)^r}{(r+1)!}D_q\left\{x^r\frac{d^{r+1}}{dx^{r+1}}f(x)\right\}$$

$$=\sum_{r=0}^{\infty}\frac{(q-1)^r}{(r+1)!}\left\{\left(\frac{d^{r+1}}{dx^{r+1}}f(x)\right)D_qx^r+(qx)^rD_q\left(\frac{d^{r+1}}{dx^{r+1}}f(x)\right)\right\}$$

By induction on n, it follows from (3.3) that:

$$D_q^nf(x)=\frac{x^{-n}q^{-n(n-1)/2}}{(q-1)^n}\sum_{j=0}^{n}\begin{Bmatrix}n\\j\end{Bmatrix}_q(-1)^j q^{j(j-1)/2}f(q^{n-j}x) \tag{3.4}$$

As a special case when $n=2$, $D_q^2f(x)$ is evaluated to give:

$$D_q^2f(x)=\frac{1}{q(q-1)^2x^2}\{f(q^2x)-(1+q)f(qx)+qf(x)\} \tag{3.5}$$

3.2. The q-Integral Operator

The q-integral operator will be defined as the inverse of the q-differential operator.

Given $d_q F(x) = f(x)$, we have:

$$F(x) - F(qx) = (1-q)xf(x) \tag{3.6}$$

It then follows that; $F(q^j x) - F(q^{j+1}x) = (1-q)q^j xf(q^j x),\ j = 0,\ldots,(n-1)$.

Hence, summing these relations over j gives:

$$F(x) - F(q^n x) = (1-q)x\sum_{r=0}^{n-1} q^r f(q^r x) \tag{3.7}$$

Assuming $q \in (0,1)$ so that $F(q^n x) \to F(0)$ as $n \to \infty$, it follows that

$$F(x) - F(0) = (1-q)x\sum_{r=0}^{\infty} q^r f(q^r x)$$

We can now formally define the q-integral of a function $f(x)$ on a given interval $[0,\alpha]$ as

$$\int_0^\alpha f(x)\mathrm{d}x_q = \alpha(1-q)\sum_{r=0}^{\infty} q^r f(q^r \alpha) = \sum_{r=0}^{\infty}(x_r - x_{r+1})f(x_r), \qquad x_r = \alpha q^r \tag{3.8}$$

On the semi-infinite interval $[\alpha,\infty)$, the q-integral of a function $f(x)$ is defined as:

$$\int_{\alpha}^{\infty} f(x)\mathrm{d}_q x = \alpha(1-q)\sum_{r=1}^{\infty} q^{-r} f\left(q^{-r}\alpha\right) \tag{3.9}$$

Over any closed interval $[\alpha,\beta]$, the q-integral of a function $f(x)$ is given by

$$\int_{\alpha}^{\beta} f(x)\mathrm{d}_q x = \int_{0}^{\beta} f(x)\mathrm{d}_q x - \int_{0}^{\alpha} f(x)\mathrm{d}_q x \tag{3.10}$$

We now define the integral over the interval $[0,\infty)$. This is achieved by setting $\alpha = 1$ in equations (3.8) and (3.9) and summing to get:

$$\int_{0}^{\infty} f(x)\mathrm{d}_q x = \int_{0}^{1} f(x)\mathrm{d}_q x + \int_{1}^{\infty} f(x)\mathrm{d}_q x = (1-q)\sum_{r=-\infty}^{\infty} q^{r} f\left(q^{r}\alpha\right) \tag{3.11}$$

The integration by parts formula of Newton-Leibnitz calculus is interpreted in the present noncommutative context as:

$$\int f_2(x) D_q f_1(x)\mathrm{d}_q x = f_1(x) f_2(x) - \int f_1(qx) D_q f_2(x)\mathrm{d}_q x$$

4. APPLICATION

There are a number of applications of the foregoing, we mention here just two, namely:

1) q-binomial formulae in two variables satisfying a quadratic relation, this has recently been published in [5] and [6] . These relations have applications in quantum group theory and non-commutative geometry.

2) A recent trend in modern physics is the study of the quantum anti-de Sitter space [7] [8] possibly in connection with q = root of unity [9] .

ACKNOWLEDGEMENTS

This work began at the African Institute for mathematical sciences (AIMS) in Muizenberg South Africa, when the first author visited in 2010. I wish to thank the director Prof Fritz Hahne for giving me a postdoctoral fellowship, and for providing a conducive environment which made this research possible.

REFERENCES

1. Connes, A. (1986) Non-Commutative Differential Geometry. Extrait des Publications Mathematiques-IHES, 62. (cited in: Qauntum Principal Bundles and Their Characteristic Classes (pdf), by MICO DURDEVIC, arXiv:q-alg/960505008vi (5 May 1996))

2. Connes A. (1994) Noncommutative Geometry. Academic Press New York.

3. Brateli O. and Robinson, D. (1979) Operator Algebras and Quantum Statistical Mechanics, Volumes 1/2. SpringerVerlag, Berlin.

4. Brown L.G., Douglas, R.G. and Filmore, P.G. (1977) Extensions of C*-Algebras and K-Homology. Annals of Mathematics, 105, 265-324.

5. Benaoum H.B. (1999) (q; h)-Analogue of Newton's Binomial Formula. Journal of Physics A: Mathematical and General, 32, 2037-2040.

6. Rosengren H. (1999) Multivariable Orthogonal Polynomials as Coupling Coefficients for Lie and Quantum Algebra Representations. Dissertation, Centre for Mathematical Sciences, Mathematics (Faculty of Science), Lund, 167.

7. Kowalski-Glikman J. (1998) Black Hole Solution of Quantum Gravity. Physics Letters A, 250, 62-66.

8. Chang Z. (1999) Quantum Anti-De Sitter Space. (reprint)

9. Steinacker H. (1998) Finite Dimensional Unitary Representations of Quantum Antide Sitter Groups at Roots of Unity. Communications in Mathematical Physics, 192, 687-706.

10

CHAPTER

Tau-Collocation Approximation Approach for Solving First and Second Order Ordinary Differential Equations

James E. Mamadu[1], Ignatius N. Njoseh[2]

[1]*Department of Mathematics, University of Ilorin, Ilorin, Nigeria*

[2]*Department of Mathematics and Computer Science, Delta State University, Abraka, Nigeria*

ABSTRACT

This paper presents Tau-collocation approximation approach for solving first and second orders ordinary differential equations. We use the method in the stimulation of numerical techniques for the approximate solution of linear initial value problems (IVP) in first and second order ordinary differential equations. The resulting numerical evidences show the method is adequate and effective.

Keywords: Ordinary Differential Equation (ODE), Initial Value Problem (IVP), Canonical Polynomial, Collocation

1. INTRODUCTION

The subject of Ordinary Differential Equation (ODE) is an important aspect of mathematics. It is useful in modeling a wide variety of physical phenomena—chemical reactions, satellite orbit, electrical networks, and so on. In many cases, the independent variable represents time so that the differential equation describes changes, with respect to time, in the system being modeled. The solution of the equation will be a representation of the state of the system. Consequently, the problem of finding the solution of a differential equation plays a significant role in scientific research, particularly, in the stimulation of physical phenomena. However, it is usually impossible to obtain direct solution of differential equations for systems to be modeled, especially complex ones encountered in real world problems. Since most of these equations are, or can be approximated by ordinary differential equations, a fast, accurate and efficient ODE solver is much needed. The Tau method was introduced by [1] to provide approximate polynomial solution for linear ordinary differential equation with polynomial coefficient.

The method takes advantage of the special properties of Chebychev polynomials. The main idea is to obtain an approximate solution of a given problem by solving an approximate problem. To further enhance the desired simplicity Lanczos introduced the systematic use of the canonical polynomials in the Tau method. The difficulties presented by the construction of such polynomials limited its application to first order ODE with the polynomial coefficient. The said difficulties were resolved by [2] when he proposed the generation of these canonical polynomials recursively. The beauty of the result of Ortiz is that the elements of canonical polynomials sequences by means of a simple recursive relation which is self starting and explicit. There are many literature developed concerning the Tau/Tau-collocation approximation method (see ([3] - [8])).

In this paper, we apply the Tau-collocation approximation method for the solution of linear initial value problems of the first and second order ODE in its differential and canonical form. We perform some numerical stimulation on some selected problems and compare the performance/effectiveness of the method with the analytic solutions given.

2. THE TAU METHOD

Lanczos [1] approximated a solution $y(x)$ of the differential

$$\sum_{r=0}^{m} P_r(x)\, y^{(r)} = f(x),\ 0 \le x \le 1 \tag{1}$$

where $P_r(x)$, $f(x)$ are polynomials. $y^{(r)}(x)$ denotes the rth order derivative of $y(x)$ with respect to x and $y^{(0)}(x)$ taken simply as $y(x)$ by a polynomial

$$y_n(x) = \sum_{r=0}^{n} a_r x^r,\ n < \infty \tag{2}$$

and determines the coefficient $a_r, r = 0(1)n$ of (2) such that $y_n(x)$ satisfies (1) perturbed by a term(s), which are calculated as part of the process. That is, $y_n(x)$ satisfies

$$\sum_{r=o}^{m} P_r(x) y^{(r)} = f(x) + H_n(x). \tag{3}$$

Then

$$H_n(x) = \sum_{r=o}^{m+s-1} {}_{m+s-r} T^*_{n-m+r+1}, \quad 0 \le x \le 1 \tag{4}$$

where m is the order of the differential equation, s is the number of over-determination,

$r = 1(1)m + s$, are the parameters to be determined, and

$$T_r^*(x) = \text{Cos}\left[r\text{Cos}^{-1}\left(\frac{2x-a-b}{b-a}\right)\right] \equiv \sum_{k=0}^{r} C_k^{(r)} x^k \tag{5}$$

is the rth degree shifted Chebychev polynomial valid in the interval $[a,b]$ (assuming (1) is defined in this interval).

The free parameters in Equation (4) and the coefficient a_r, $r \in 0(1)n$ in (2) are obtained by equating the values of x in (3) together with (1) to zero.

2.1. Description of the Differential Form

Considering the mth order linear differential Equation ([1] [2])

$$L(y(x)) \coloneqq \sum_{r=o}^{m} P_r(x) y^{(r)} = f(x), \tag{6}$$

$$y(0)=\alpha_0, y'(0)=\alpha_1,\cdots,y^{(m-1)}(0)=\alpha_{(m-1)},$$

with y(x) as the exact solution in $a \le x \le b, |a| < \infty, |b| < \infty.$

We seek an approximate solution of the differential solution by the Tau method using the nth degree polynomial function

$$y_n(x)=\sum_{r=o}^{m} P_r(x)x^r, n<\infty \tag{7}$$

which satisfies the perturbed problem

$$L(y(x)):=\sum_{r=o}^{m} P_r(x)y^{(r)}=f(x)+H_n(x) \tag{8}$$

We equate the corresponding coefficient of x in (8) and using the initial conditions

$$y(0)=\alpha_0, y'(0)=\alpha_1,\cdots,y^{(m-1)}(0)=\alpha_{(m-1)},$$

We then solve the system of equation by Gaussian elimination method.

2.2. Collocation Approach to the Tau Method

The Lanczos Tau method in [4] is an economization process for a function that is implicitly defined by differential equation. Let us assume an approximation of the power series expansion $u(x)$ as

$$u_n(x)=\sum_{r=o}^{n}a_r(x)x^r \tag{9}$$

Consider an approximation to the residual $R_n(x)$ as

$$R_n(x)={}_1T_n^*(x)+{}_2T_{n-1}^*(x)+\cdots+{}_mT_{n-m+1}^*(x) \tag{10}$$

Then by the Tau method, if

$$L(u(x))=f(x) \tag{11}$$

we have

$$L(u(x))=f(x)+R_n(x)=f(x)+{}_1T_n^*(x)+{}_2T_{n-1}^*(x)+\cdots+{}_mT_{n-m+1}^*(x) \tag{12}$$

where L is a linear differential operator of order n.

We collocate (12) at $x_k=kh,k=1(1)N,$ where $h=\dfrac{b-a}{n+1}$ to have

$$L(u(x_k))=f(x_k)+R_n(x_k)=f(x_k)+{}_1T_n^*(x_k)+{}_2T_{n-1}^*(x_k)+\cdots+{}_mT_{n-m+1}^*(x_k) \tag{13}$$

The parameter $1,2,\cdots,m$ may be eliminated leaving the unknown coefficient $a_k,k=0(1)n$ with $(n+1)$ linear equations which can be solved by Gaussian elimination.

3. ERROR ESTIMATION

Let us in this section consider and obtain the error estimator for the approximate solution of (1) and (9). Let $E_n = u(x) - u_n(x)$ be the error function of $u_n(x)$ to $u(x)$, where $u(x)$ is the exact solution of (1) and (9). Therefore $u_n(x)$ satisfies the perturbed problems:

$$L(u_n(x)) := \sum_{r=0}^{m} P_r(x)(u_n(x))^{(r)} = f(x) + H_n(x), x \in [a,b], |a| < \infty, |b| < \infty, n < \infty \tag{14}$$

and

$$L(u_n(x)) := \sum_{r=0}^{m} a_r x^r = f(x) + H_n(x) \tag{15}$$

where H_n is uniquely defined as in (4).

To obtain the perturbation term $H_n(x)$, we substitute the computed solution $u_n(x)$ such that

and

$$L(u_n(x)) := \sum_{r=0}^{m} a_r x^r - f(x) = H_n(x)$$

We then proceed to find an approximate $E_{n,N}(x)$ to the error function $E_n(x)$ in the same manner as we did for the solution of (1) and (9).

Thus, the error function, E_n, satisfy the problem

$$L(E_n) := \sum_{r=0}^{m} P_r(x) E_n^r - f(x) = -H_n(x) \tag{16}$$

and

$$L(E_n) := \sum_{r=0}^{m} a_r x^r - f(x) = 0 \tag{17}$$

which satisfies the conditions prescribed.

4. ILLUSTRATIVE EXAMPLES

In this section, two initial value problems are considered to show the efficiency of the method.

Example 1

Consider linear initial value problem in second order ordinary differential equation

$$\begin{cases} u''(x) + u(x) = 0 \\ u(0) = 1, u'(1) = 3 \end{cases} \tag{18}$$

We solve [4] for $n = 2$ using; (i) The Tau method; and (ii) Tau-collocation method.

The analytic solution is $u(x) = \text{Cos}x - \dfrac{\text{Sin}x(\text{Cos}1 - 3)}{\text{Sin}1}$.

By the Tau method we obtain the linear differential operator as

$$L \equiv \left(\frac{d^2}{dx^2} + 1 \right) \tag{19}$$

The associated canonical polynomials are obtained as follows:

$$Lx^r = \left(\frac{d^2}{dx^2} + 1 \right) x^r \tag{20}$$

$$Lx^r = r(r-1)LQ_{r-2}(x) + LQ_r(x)$$

$$L^{-1} \cdot Lx^r = L^{-1} \cdot L\left[r(r-1)Q_{r-2}(x) + Q_r(x) \right]$$

$$\Rightarrow x^r = r(r-1)Q_{r-2}(x) + Q_r(x)$$

$$\Rightarrow Q_r(x) = x^r - r(r-1)Q_{r-2}(x)$$

The canonical polynomials, $Q_r(x), r \geq 0$, obtained here can easily be obtained from [3] where the generalized form of the canonical polynomials was reported.

For $r = 0(1)2$, we have

$Q_0(x) = 1, Q_1(x) = x$ and $Q_2(x) = x^2 - 2$

These polynomials are substituted into Equation (12) to give

$$Lu_n(x) = {}_1T_n^*(x) + {}_2T_{n-1}^*(x) \tag{21}$$

Using Equation (5),

$$Lu_n(x) = {}_1\sum_{r=0}^{n} C_r^{(n)} x^r + {}_2\sum_{r=0}^{n-1} C_r^{(n-1)} x^r(x)$$

$$Lu_n(x) = {}_1\sum_{r=0}^{n} C_r^{(n)} LQ_r(x) + {}_2\sum_{r=0}^{n-1} C_r^{(n-1)} LQ_r(x)$$

$$u_n(x) = {}_1\sum_{r=0}^{n} C_r^{(n)} Q_r(x) + {}_2\sum_{r=0}^{n-1} C_r^{(n-1)} Q_r(x) \tag{22}$$

Since

$$u_2(x) = {}_1\sum_{r=0}^{2} C_r^{(2)} Q_r(x) + {}_2\sum_{r=0}^{1} C_r^{(n-1)} Q_r(x)$$

Now,

$T_0(x) = 1, T_1(x) = 2x - 1$ and $T_2(x) = 8x^2 - 8x + 1$

$$u_2(x) = {}_1\left(C_0^{(2)}Q_0(x) + C_1^{(2)}Q_1(x) + C_2^{(2)}Q_2(x)\right) + {}_2\left(C_0^{(1)}Q_0(x) + C_1^{(1)}Q_1(x)\right)$$

$$\Rightarrow u_2(x) = {}_1\left(1 - 8x + 8\left(x^2 - 2\right)\right) + {}_2(-1 + 2x) \tag{23}$$

Using initial conditions on Equation (23) and simplifying further we get the approximate solution as

$$u_2(x) = 1 + \frac{46}{15}x - \frac{16}{15}x^2.$$

Considering the Tau-collocation method we have:

Let

$$u_2(x) = \sum_{r=0}^{2} a_r(x)x^r \tag{24}$$

$$u_2(x) = a_0 + a_1x + a_2x^2$$

$$u_2'(x) = a_1 + 2a_2x$$

$$u_2''(x) = 2a_2$$

Substituting into (13) we have,

$$2a_2 + a_0 + a_1x + a_2x^2 = {}_1T_2^*(x) + {}_2T_1^*(x) \tag{25}$$

$$2a_2 + a_0 + a_1x + a_2x^2 = {}_1\left(1 - 8x + 8\left(x^2 - 2\right)\right) + {}_2(-1 + 2x)$$

Now collocating at $x_k = kh, h = \dfrac{b-a}{n+1}, k = 1,2,3$ and using the initial conditions, we obtain the approximate solution as

$$u_2(x) = 1 + \frac{46}{15}x - \frac{16}{15}x^2$$

Example 2

Consider the first order IVP

$$Ly(x) = (1+x)y'(x) + y(x) = 0, \; y(0) = 1 \tag{26}$$

The exact solution is $y(x)=\frac{1}{x+1}$

For the given IVP, we can deduce that $m=1$ and $s=0$.

The differential formulation is as follows:

Let

$$y_n(x)=\sum_{r=o}^{n}a_r x^r \tag{27}$$

Taking $n=5$

$$Ly_5(x)=\sum_{r=o}^{m+s-1}{}_{m+s-r}T^*_{n-m+r+1}(x)$$

$$\Rightarrow Ly_5(x)=\sum_{r=o}^{0}{}_1T^*_5$$

where

$$T^*_5=\sum_{r=o}^{5}C_r^{(5)}x^r \tag{28}$$

hence

$$(1+x)y_5'(x)+y_5(x)={}_1T^*_5(x) \tag{29}$$

but

$$Ly_5(x)={}_1\sum_{r=o}^{5}C_r^{(5)}x^r$$

Using (28) and (30) in (29) we obtain,

$$(x+1)\sum_{r=o}^{5} ra_r x^{r-1} + \sum_{r=o}^{5} ra_r x^r = {}_1\sum_{r=o}^{5} C_r^{(5)} x^r \tag{31}$$

Expanding and equating coefficients of powers of x, the resulting linear equations together with the equations obtained using the initial conditions is written in the form,

$$A\underline{x} = \underline{b}$$

where

$$\begin{pmatrix} 1 & 1 & 0 & 0 & 0 & 0 & \cdots & C_0^{(5)} \\ 0 & 2 & 2 & 0 & 0 & 0 & \cdots & C_1^{(5)} \\ 0 & 0 & 3 & 3 & 0 & 0 & \cdots & C_2^{(5)} \\ 0 & 0 & 0 & 4 & 4 & 0 & \cdots & C_3^{(5)} \\ 0 & 0 & 0 & 0 & 5 & 5 & \cdots & C_4^{(5)} \\ 0 & 0 & 0 & 0 & 0 & 0 & \cdots & C_5^{(5)} \\ 1 & 0 & 0 & 0 & 0 & 0 & \cdots & 0 \end{pmatrix}$$

$$\underline{x} = (a_0, a_1, a_2, a_3, a_4, a_5, 1)^T,$$

$$\underline{b} = (0,0,0,0,0,0,1)^T$$

Using Equation (5), we obtain the following values,

$C_0^{(5)} = -1, C_1^{(5)} = 50, C_2^{(5)} = -400, C_3^{(5)} = 1120, C_4^{(5)} = -1280$ and $C_5^{(5)} = 512.$

Using these values in the matrix and solving by Gaussian elimination method, we have,

$$a_0 = 1, a_1 = -\frac{2339}{2342}, a_2 = \frac{1132}{1171}, a_3 = -\frac{932}{1171}, a_4 = \frac{512}{1171}, a_5 = -\frac{128}{1171}, {}_1 = \frac{3}{2342}$$

The approximate solution is:

$$y_5(x) = \frac{1}{2342}\left(2342 - 2339x + 2264x^2 - 1864x^3 + 1024x^4 - 256x^5\right)$$

DISCUSSION OF RESULTS

The results obtained above show that the Tau-collocation method is appropriate for the solution of linear initial value problems of first and second kind ordinary differential equations. From the tables (Table 1 and Table 2) of results presented above, we observe that the approximate solution considered at grid points, $n = 2$ and $n = 5$, for examples 1 and 2 converges to the

Table 1. Numerical results for example 1.

X	*Exact Solution*	*Approximate solution, n = 2*	*Error*
0.1	1.2868265	1.2960000	9.1735e−03
0.2	1.5607954	1.5706667	9.8712e−03
0.3	1.8191694	1.8240000	4.8306e−03
0.4	2.0593669	2.0560000	3.3669e−03
0.5	2.2789879	2.2666667	1.2321e−02
0.6	2.4758379	2.4560000	1.9838e−02
0.7	2.6479502	2.6240000	2.3950e−02
0.8	2.7936051	2.7706667	2.2938e−02
0.9	2.9113471	2.8960000	1.5347e−02
1.0	3.0000000	3.0000000	0.0000e+00

Table 2. Numerical results for Example 2.

X	*Exact Solution*	*Approximate solution, n = 2*	*Error*
0.1	0.9090909	0.9090418	4.19133e-05
0.2	0.8333333	0.8332214	1.1195e-04
0.3	0.7692308	0.7691735	5.7276e-05
0.4	0.7142857	0.7143198	3.4081e-05
0.5	0.6666667	0.6667378	7.1164e-05
0.6	0.6250000	0.6250298	2.9821e-05
0.7	0.5882353	0.5881915	4.3800e-05
0.8	0.5555556	0.5554809	7.4633e-05
0.9	1.5263158	0.5265873	2.8445e-05
0.10	0.5000000	0.5000000	0.0000e+00

analytic solution with maximum absolute errors of 10^{-3} and 10^{-5} respectively. We obtain satisfactory results because of the excellent convergence rate of the Tau-collocation approximation method, which is very close to the minimax polynomial which minimizes the maximum error in approximation. Thus, the approximate solution will match the analytic solution as n increases.

5. CONCLUSIONS

This paper has considered Tau-collocation approximation approach for solving particular first and second order ordinary differential equations. The method offers several advantages which include, among others:

1) It takes advantages of the special properties of Chebychev polynomials which can be easily generated recursively;

2) Elements of canonical polynomials sequences by means of a simple re-cursive relation which is self starting and explicit; and

3) It can easily be programmed for experimentation.

Tau-Collocation method can be extended to higher order ordinary differential equations and stochastic differential equations. It can also be used to solve integro-differential and stochastic integro-differential equations.

REFERENCES

1. Sam, C.N. (2004) Numerical Solution of Partial Differential Equations with the Tau-Collocation Method. M. Phil Thesis, City University of Hong Kong.
2. Khajah, H. (1999) Tau-Method Approximation of a Generalized Epstein-Hubbel Elliptic Type Integral. Computer and Mathematics, 68, 1615-1621.
3. Ortiz, E.L. (1975) Step by Step Tau-Method—Part 1. Computer and Mathematics with Applications, 1, 381-392.
4. El-Daou, M.K., Ortiz, E.L. and Samara, H. (1993) A Unified Approach to the Tau Method and Chebychev Series Expansion Technique. Computer and Mathematics with Applications, 25, 73-82.
5. Coleman, J.P. (1974) Lanczos Tau Method. IMA Journal of Applied Mathematics, 7, 85-97.
6. Yisa, B.M. and Adeniyi, R.B. (2012) A Generalized Formulation for Canonical Polynomials for M-Th Order Non Over-Determined Ordinary Differential Equations. Internal Journal of Engineering Research and Technology.

7. Ortiz, E.L. (1969) The Tau Method. SIAM Journal on Numerical Analysis, 6, 480-492.

8. Lanczos, C. (1956) Applied Analysis. Prentice-Hall, Engle-Wood Cliffs, New Jersey.

11

CHAPTER

The Continuous Analogy of Newton's Method for Solving a System of Linear Algebraic Equations

Tugal Zhanlav[1], Ochbadrakh Chuluunbaatar[1,2*], Gantumur Ankhbayar[1]

[1]*School of Mathematics and Computer Science, National University of Mongolia, Ulan-Bator, Mongolia*

[2]*Joint Institute for Nuclear Research, Dubna, Moscow Region, Russia*

ABSTRACT

We propose a continuous analogy of Newton's method with inner iteration for solving a system of linear algebraic equations. Implementation of inner iterations is carried out in two ways. The former is to fix the number of inner iterations in advance. The latter is to use the inexact Newton method for solution of the linear system of equations that arises at each stage of outer iterations. We give some new choices of iteration parameter and of forcing term, that ensure the convergence of iterations. The performance and efficiency of the proposed iteration is illustrated by numerical examples that represent a wide range of typical systems.

Keywords: Continuous Analogy of Newton's Method; Solving the System of Linear Algebraic Equations; Convergence; Choice of Iteration Parameter

1. INTRODUCTION

We consider a system of linear algebraic equations

$$Ax = f. \tag{1}$$

For a numerical solving of Equation (1) we consider an iterative process:

$$Bv_n = -r_n, x^{(n+1)} = x^{(n)} + \tau_n v_n, n = 0,1,\cdots \tag{2}$$

Of course, the quality of iterative process Equation (2) essentially depends on the choices of matrix B and of iteration parameter $\tau_n \neq 0.$

Let H be a linear space of m-dimensional vectors. We will denote the Euclidean vector norm in H by $\|\cdot\|$, as well as the corresponding norm of matrices.

Theorem 1.1. Let $(Av_n, r_n) \neq 0.$ Then a necessary and sufficient condition for the $\|r_n\|$ to be decreasing $(\|r_{n+1}\| < \|r_n\|)$ is that

$$\tau_n \in I_n = \begin{cases} \left(0, -\dfrac{2(Av_n, r_n)}{\|Av_n\|^2}\right), & \text{when } (Av_n, r_n) < 0, \\ \left(-\dfrac{2(Av_n, r_n)}{\|Av_n\|^2}, 0\right), & \text{when } (Av_n, r_n) > 0. \end{cases}$$

Proof. From Equation (2) we obtain

$$r_{n+1} = r_n + \tau_n A v_n.$$

Hence we have

$$\| r_{n+1} \|^2 = \| r_n \|^2 + \tau_n [\tau_n \| Av_n \|^2 + 2(Av_n, r_n)]. \qquad (3)$$

The assertion follows from (3).

The interval I_n we call τ-region of convergence of the iteration method (2). Thus we have to choose τ_n from this region. Moreover, it is desirable that the τ_n to be optimal in some sense. Further we will use wellknown assertions to study the convergence of (2).

Theorem 1.2. [1]. Let S be an m × m matrix. Then the successive approximations

$$x^{(n+1)} = Sx^{(n)} + z, \; n = 0, 1, 2, \cdots \qquad (4)$$

converge for each $z \in R^m$ and each $x^{(0)} \in R^m$ if and only if

$$\rho(S) < 1, \tag{5}$$

where $\rho(S)$ is a spectral radius of the matrix S.

It is easy to show that the iteration process (2) can be rewritten as (4) with iteration matrix

$$S = E - \tau_n B^{-1} A \text{ and } z = \tau_n B^{-1} f. \tag{6}$$

Here E is an m × m unit matrix.

Theorem 1.3. [2]. The iteration process (2) with parameter τ_n given by

$$\tau_n^* = -\frac{(Av_n, r_n)}{\|Av_n\|^2} \tag{7}$$

converges to x^* for any $x^{(0)} \in H$. The following a posteriori estimate holds true:

$$\|r_n\| = q_{n-1} q_{n-2} \cdots q_0 \|r_0\|, \tag{8}$$

where

$$q_i = \sqrt{1 - \frac{(Av_i, r_i)^2}{\|Av_i\|^2 \|r_i\|^2}} < 1, \ i = 0, 1, \cdots, n-1.$$

We call the nonzero value of τ_n^* defined by (10) the optimal one in the sense that it yields the minimum value of functional $\|r_{n+1}\|$.

2. THE CONTINUOUS ANALOGY OF NEWTON'S METHOD

The continuous analogy of Newton's method is also applicable to (1) and leads to [2]

$$Av_n = -r_n, \tag{9a}$$

$$x^{(n+1)} = x^{(n)} + \tau_n v_n, n = 0,1,\cdots \tag{9b}$$

It should be mentioned that not only the convergence, but also the convergence rate of iteration (9) depends on the choice of the parameter τ_n. We have the following:

Lemma 2.1. The sufficient condition for $\|r_n\|$ to be decreasing is $0 < \tau_n < 2, n = 0,1,\cdots$

Lemma 2.2. Suppose that $1 < \tau_n < 2$ for all $n \geq 0$. Then the iteration process (2) gives two-sided approximations to x^*, i.e.,

$$x^{(1)} < \cdots < x^{(2m+1)} < x^* < x^{(2m)} < \cdots < x^{(0)} \text{ if } x^* < x^{(0)}$$

or

$$x^{(0)} < \cdots < x^{(2m)} < x^* < x^{(2m+1)} < \cdots < x^{(1)} \text{ if } x^{(0)} < x^*.$$

The proofs are immediately followed from the equalities

$$r_{n+1} = (1-\tau_n) r_n,\ x^{(n+1)} - x^* = (1-\tau_n)\left(x^{(n)} - x^*\right).$$

At each step of iterations one can solve the system (9a) by means of some iterative methods. We call this inner iteration. We consider the following decomposition of A:

$$A = A_1 + A_2,$$

in which the matrix A_1 is simple and invertible. For finding the correction v_n we use inner iteration

$$A_1 v_n^{(e)} = -r_n - A_2 v_n^{(e-1)},\ v_n^{-1} \equiv 0,\ e = 0,1,\cdots \tag{10}$$

Theorem 2.3. Suppose that

$$\|C\| < 1,\ C = A_2 A_1^{-1}. \tag{11}$$

Then the inner iteration (10) converges and holds the following estimate

$$\left\|v_n - v_n^{(k)}\right\| \le \left\|A_1^{-1}\right\| \frac{\|C\|^{k+1}}{1-\|C\|} \|r_n\|, k = 0,1,\cdots \tag{12}$$

Proof. The linear system (10) can be rewritten as

$$(E+C) A_1 v_n = -r_n. \tag{13}$$

By assumption (11) there exists $(E+C)^{-1}$ and the following series representation is valid

$$(E+C)^{-1} = \sum_{j=0}^{\infty}(-1)^j C^j. \tag{14}$$

Then from (13) it follows that

$$v_n = -A_1^{-1}\left(\sum_{j=0}^{\infty}(-1)^j C^j\right)r_n. \tag{15}$$

In a similar way, from (10) we have

$$v_n^{(k)} = -A_1^{-1}\left(\sum_{j=0}^{k}(-1)^j C^j\right)r_n. \tag{16}$$

From (15) and (16) immediately follows (14).

The estimate (12) means that the inner iteration (10) converges under condition (11). In real computations we have to terminate the inner iteration before convergence. We will restrict the number of inner iteration by k. Then the whole iteration process looks like:

$$A_1 v_n^{(l)} = -r_n - A_2 v_n^{(l-1)},\ v_n^{-1} \equiv 0, l = 0,1,\cdots,k, \tag{17a}$$

$$x^{(n+1)} = x^{(n)} + \tau_n v_n^{(k)},\ n = 0,1\cdots \tag{17b}$$

We now formulate the convergence theorems for these methods.

Theorem 2.4. Suppose that the condition (11) is satisfied and the iteration parameter τ_n is given by

$$\tau_n = -\frac{\left(Av_n^{(k)}, r_n\right)}{\left\|Av_n^{(k)}\right\|^2}. \tag{18}$$

Then the iteration process (17) converges for any $k = 0,1,\cdots$ and for arbitrary chosen starting $x^{(0)}$.

Proof. We rewrite the iteration process (17) as (2) with

$$B = \left(\sum_{j=0}^{k} (-1)^j C^j\right)^{-1} A_1.$$

By Theorem 1.3, such a process converges if we choose τ_n by (7), in which the v_n is replaced by $v_n^{(k)}$.

Obviously the number k may be different for each n. From (12) it is evident, that it suffices to restrict the number of inner iteration only by k = 0 or 1, when the residual norm $\|r_n\|$ is small enough.

Theorem 2.5. Suppose that the condition (11) is satisfied. Then the iteration process (17) with $\tau_n \in (0,1]$ converges for any $k = 0,1,\cdots$ and for arbitrary chosen starting $x^{(0)}$. The following inequality holds:

$$\|r_{n+1}\| \le \theta_n \|r_n\|, \quad 0 < \theta_n < 1. \tag{19}$$

Proof. From (17) we get

$$r_{n+1} = r_n + \tau_n A v_n^{(k)}.$$

Using (11) and (16) we rewrite the last expression as

$$r_{n+1} = \left((1-\tau_n)E + \tau_n (-1)^{k+1} C^{k+1}\right) r_n. \tag{20}$$

If $0 < \tau_n \le 1$, then from (20) we obtain

$$\|r_{n+1}\| \le \theta_n \|r_n\|,\ \theta_n = 1 - \tau_n\left(1 - \|C\|^{k+1}\right).$$

According to (11), we have $0 < \theta_n < 1$. The convergence of (17) follows from (19).

Corollary 2.6. Let the condition (11) fulfill, and τ_n is given by [3]

$$\tau_n = \min\left\{\frac{\|r_{n-1}\|}{\|r_n\|}\tau_{n-1};1\right\},\ n = 1,2,\cdots,\ \tau_0 \approx 0.1.$$

Then the iteration (17) converges.

Remark 2.7. In proofs of Theorems 2.3-2.5 the condition (11) is essentially used. In particular, it may be fulfilled if the matrix A of the system (1) is a strongly diagonally dominant and A_1 is chosen as

$$A_1 = \operatorname{diag}\{a_{11},\cdots,a_{mm}\}.$$

Theorem 2.8. Suppose that the condition (11) is satisfied and that the eigenvalues of matrix $D = A_1^{-1}A_2$ are real. Then the spectral radius of iteration matrices of the iterations (17) is decreasing with respect to k when $\tau_n \to 1$.

Proof. The iterations (17) can be rewritten as (4) with iteration matrices

$$S_k = (1-\tau_n)E + \tau_n(-1)^{k+1} D^{k+1}. \tag{21}$$

We denote by $\lambda(D)$ the eigenvalues of D. If $\tau_n = 1$, then

$$S_k = (-1)^{k+1} D^{k+1}.$$

Therefore, we have

$$|\lambda(S_k)| = |\lambda(D)|^{k+1}.$$

By virtue of (11) we obtain $|\lambda(C)| \le \|C\| < 1$. Since the spectrums of matrices C and D coincide, we also have

$$|\lambda(D)| < 1.$$

By definition we get

$$\begin{aligned}\rho(S_k) &= \max|\lambda(S_k)| = (\max|\lambda(D)|)^{k+1}\\ &= (\max|\lambda(D)|)^{k} \max|\lambda(D)|\\ &< \max|\lambda(D)|^{k} = \rho(S_{k-1}),\end{aligned}$$

which is valid for all $k = 1, 2, \cdots$ Thus we have

$$\rho(S_k) < \rho(S_{k-1}) < \cdots < \rho(S_1) < \rho(S_0) < 1. \tag{22}$$

Obviously, from (21) it is clear that S_k is a continuous function of τ_n. Therefore, (22) is valid for $\tau_n \to 1$. The iteration (17) with a few small k represents a special interest from a computational view point. Moreover, it is worth to stay at (17) with k = 0 in detail. The iteration (17) with k = 0 and $A_1 = \text{diag}\{a_{11}, \cdots, a_{mm}\}$ leads to the well-known Jacobi iteration with relaxation parameter $\omega = \tau_n$. It is also known that [1] the Jacobi method with optimal relaxation parameter

$$\omega_{opt} = \frac{2}{2 - \lambda_{\min} - \lambda_{\max}} \tag{23}$$

converges under the assumption that the Jacobi matrix $B = -A_1^{-1} A_2$ has real eigenvalues and spectral radius less than one. Here by $\lambda_{\min}$ and $\lambda_{\max}$ denoted are the smallest and largest eigenvalues of B. Fortunately, we can prove the convergence of Jacobi method with relation parameter under a mild condition than the above mentioned assumption. Namely, we have.

Corollary 2.9. Suppose that the condition (11) is satisfied. Then the Jacobi method with relaxation parameter:

$$\omega_n = -\frac{\left(Av_n^{(0)}, r_n\right)}{\left\|Av_n^{(0)}\right\|} = \frac{\left((E+C)r_n, r_n\right)}{\left\|(E+C)r_n\right\|^2} \tag{24}$$

converges for any starting.

From (24) it is clear that the relaxation parameter changes depending on n. Therefore the iteration (17) with k = 0, and $A_1 = \operatorname{diag}\{a_{11},\cdots,a_{mm}\}$ and with ω_n given by (24) we call nonstationary Jacobi method with optimal relaxation parameters. The iteration (17) with k = 0 and with a lower triangular matrix A_1 leads to GaussSeidel with one parameter.

The Formula (18) can be rewritten as:

$$\tau_n = \frac{\left(AB^{-1}r_n, r_n\right)}{\left\|AB^{-1}r_n\right\|^2}, \tag{25}$$

where $AB^{-1} = E - (-1)^{k+1} C^{k+1}$. From this it is clear that $\tau_n \to 1,$ as $k \to \infty$. This means that whole iteration (17) with the parameter given by (18) converges quadratically in the limit.

Since $\tau_n \to 1,$ as $k \to \infty$, then τ-region of convergence for iterations (17) leads to

$$I_n^k = \left(0; \frac{2\left(Av_n^{(k)}, r_n\right)}{\left\|Av_n^{(k)}\right\|^2}\right) \to (0;2).$$

The number of outer iteration n depends on the number of inner iteration k, i.e., $n = n_k$. In general, n is a decreasing function of k, i.e., $n_{k+1} < n_k$.

On the other hand, the iteration (17) can be considered as a defect correction iteration [1]

$$x^{(n+1)} = x^{(n)} - \tau_n A_{\text{approx}}^{-1} r_n, \left(B^{-1} = A_{\text{approx}}^{-1}\right), \tag{26}$$

where A_{approx}^{-1} defined by

$$A_{\text{approx}}^{-1} = A_1^{-1}\left(E - C + C^2 + \cdots + (-1)^k C^k\right). \tag{27}$$

A_{approx}^{-1} is a reasonable approximation to A^{-1} since

$$\left\| A^{-1} - A_{\text{approx}}^{-1} \right\| \leq \frac{\left\|A_1^{-1}\right\| \left\|C\right\|^{k+1}}{1 - \left\|C\right\|} \to 0 \tag{28}$$

for large k. The choice of parameter τ_n given by (25) allows us to decrease the residual norm from iteration to iteration. By this reason we call (26) as a minimal defect (or residual) correction iteration.

From (26) it follows that

$$r_{n+1} = \left(E - \tau_n A A_{\text{approx}}^{-1}\right) r_n. \tag{29}$$

From (29) and $A = A_1 + A_2 = (E + C) A_1, C = A_2 A_1^{-1}$, it follows that

$$E - \tau_n A A_{\text{approx}}^{-1} = (1 - \tau_n) E + \tau_n (-1)^{k+1} C^{k+1}.$$

Therefore we have

$$\lambda\left(E-\tau_n AA_{\text{approx}}^{-1}\right)=1-\tau_n+\tau_n(-1)^{k+1}\left[\lambda(C)\right]^{k+1}.$$

This means that the spectral radius of matrix $E-\tau_n AA_{\text{approx}}^{-1}$ depends on the number of inner iteration k, i.e., $\rho=\rho(k)$. Therefore we have $\rho(k)\to 0$ as $k\to\infty$, because $\tau_n\to 1$ as $k\to\infty$ and from (11)

$$\left|\lambda(C)\right|\le\|C\|<1.$$

Above we considered the cases, where the inner iteration number k is fixed at each outer iteration. It is desirable that the termination criterion for the inner iteration must be chosen carefully to preserve the superlinear convergence rate of the outer iteration. We stay at this problem in more detail. When $\tau\equiv 1$, the iteration (17) is, indeed, inexact Newton (IN) method for (1). Therefore, iteration (17) with parameter τ_n given by formula (18) we call inexact damped Newton method (IDN). According to IN method [4,5], we must choose $\eta_n\in[0,1)$ and continue the inner iteration until satisfy the condition

$$\left\|r_n^{(e)}\right\|\le\eta_{n-1}\left\|r_n\right\|,\ r_n^e=Av_n^{(e)}+r_n,\ n=1,2,\cdots \tag{30}$$

Thus (30) is a stopping criterion for inner iteration (17a). There are several choices of forcing term η_n in inexact Newton method [4-6]. For examples, in [6] two choices of η_n were proposed:

$$\eta_n = \begin{cases} 1-\eta_{n-1}\alpha_n, & \text{when } \eta_{n-1}\alpha_n < 1, \\ -\dfrac{1-\eta_{n-1}\alpha_n}{\alpha_n}, & \text{when } \eta_{n-1}\alpha_n \geq 1, \end{cases} \tag{31}$$

with

$$\alpha_n = \frac{\|r_{n-1}\|}{\|r_n\|},$$

and

$$\eta_n = |1-\tau_n|, \ n = 0,1,\cdots \tag{32}$$

Since we have formula (18) for τ_n, one can use the second choice (14) with no additional calculations. We can also use formula for $\tau_n = \dfrac{2}{1+\sqrt{1+\|r_n\|}}$ [7]. Therefore, according to (32), we get

$$\eta_n = \frac{\sqrt{1+\|r_n\|}-1}{\sqrt{1+\|r_n\|}+1}. \tag{33}$$

3. NUMERICAL RESULTS

The quality of the proposed iteration was checked up for numerous examples. We express the matrix A as

$$A = D + A_L + A_U,$$

where $D = \mathrm{diag}\{a_{11}, \cdots, a_{mm}\}$, and A_L, A_U are lower and upper triangular matrices, respectively. All examples are calculated with an accuracy $\|Ax_n - f\| < 10^{-7}$.

The numerical calculations are performed on Acer, CPU 1.8 GHz, 1 GB RAM, and using a software MATLAB R2007a for the Windows XP system.

Example 1. We consider a system of Equation (1) with matrix A and f given by

$$A = \begin{pmatrix} 2 & 1 & & & & \\ 1 & 4 & 1 & & & \\ & & \ddots & \ddots & \ddots & \\ & & & 1 & 4 & 1 \\ & & & & 1 & 2 \end{pmatrix}_{m\times m}, f = 6\times \begin{pmatrix} 0.5 \\ 1.0 \\ \vdots \\ 1.0 \\ 0.5 \end{pmatrix}_m,$$

which was solved by the proposed iteration method (17), (18). For comparison, it was solved by Jacobi and successive over relaxation (SOR) iterations with a parameter ω_{opt}.

Example 2.

$$A = \begin{pmatrix} 1.1161 & 0.1254 & 0.1397 & 0.1490 \\ 0.1582 & 1.1675 & 0.1768 & 0.1871 \\ 0.1968 & 0.2071 & 1.2168 & 0.2271 \\ 0.2368 & 0.2471 & 0.2568 & 1.2671 \end{pmatrix}, f = \begin{pmatrix} 1.5471 \\ 1.6471 \\ 1.7471 \\ 1.8471 \end{pmatrix}.$$

Example 3.

$$A=\begin{pmatrix} 0.60 & -0.16 & -0.12 & -0.07 & -0.03 \\ -0.16 & 0.74 & -0.31 & -0.19 & -0.07 \\ -0.12 & -0.31 & 0.66 & -0.31 & -0.12 \\ -0.07 & -0.19 & -0.31 & 0.74 & -0.16 \\ -0.03 & -0.07 & -0.12 & -0.16 & 0.96 \end{pmatrix}, f=\begin{pmatrix} 1.0 \\ 0.1 \\ 0.1 \\ 0.1 \\ 1.0 \end{pmatrix}.$$

Example 4.

$$A=\begin{pmatrix} T_N+2I_N & -I_N & & & \\ -I_N & T_N+2I_N & -I_N & & \\ & \ddots & \ddots & & \\ & & -I_N & T_N+2I_N & -I_N \\ & & & -I_N & T_N+2I_N \end{pmatrix}_{N^2\times N^2},$$

$$f=h^2\begin{pmatrix} 1 \\ 1 \\ \vdots \\ 1 \\ 1 \end{pmatrix}_{N^2},$$

where $h=1/(N+1)$ and

$$T_N=\begin{pmatrix} 2 & -1 & & & \\ -1 & 2 & -1 & & \\ & \ddots & \ddots & & \\ & & -1 & 2 & -1 \\ & & & -1 & 2 \end{pmatrix}_{N\times N}, I_N=\begin{pmatrix} 1 & & & & \\ & 1 & & & \\ & & \ddots & & \\ & & & 1 & \\ & & & & 1 \end{pmatrix}_{N\times N}.$$

Such a system arises in discretization of two dimensional Poisson equation [8]:

$$\Delta u(x,y) = -1,\ D = \{(x,y) | 0 \le x, y \le 1\},\ u = 0 \text{ on } \partial D.$$

The exact value of $u(1/2,1/2)$ is [9]

$$u\left(\frac{1}{2},\frac{1}{2}\right)$$

$$= \frac{16}{\pi^4} \sum_{\mu,\nu=0}^{\infty} \frac{(-1)^{\mu+\nu}}{(1+2\mu)(1+2\nu)\left((1+2\mu)^2 + (1+2\nu)^2\right)}$$

$$= 0.0736713 \cdots \quad (34)$$

The properties of matrix for examples 1-4 are shown in **Table 1**. From this we see that the considered examples represent a wide range of typical systems.

The numbers of Jacobi and SOR iterations versus the dimension m of example 1 are presented in**Table 2**. Here k is the number of inner iteration in (17) (18). From this example we see that the proposed iteration (17), (18) can compete with SOR iteration with optimal relaxation parameter and seem to be superior to the Jacobi iteration. The behavior of the iteration parameter τ_n given by (18) at m = 10 is explained in **Table 3**. From this we see that the iteration parameter τ_n tends to 1 as k increases.

The number n of outer iteration (17), (18) with fixed number k of inner iterations, and CPU time for examples 2 and 3 are shown in **Table 4**. From this we see that they are in example 2 considerable less than example 3. This explained by reason that matrix of this system has a strictly diagonally dominant. The number n of outer iteration, the total number k of inner iteration when forcing terms η_n was chosen by formulas (32) and (33), and CPU time are displayed in **Table 5**. Here it is observed similar situations as in the previous case.

Monotonic convergence of the calculated values $u^h(1/2,1/2)$ to exact value (34) versus the dimension N of the example 4 is shown in **Table 6**. The number n of outer iteration with fixed and unfixed number k of inner iteration and CPU time versus the dimension N of the example 4 are presented in Tables 7 and 8, respectively.

Table 1. The properties of matrix for examples 1-4.

Example	Matrix		
	Symmetric	Diagonally dominant	Sparse
1	+	+	+
2	–	+	–
3	+	–	–
4	+	–	+

Table 2. The numbers of Jacobi and SOR iterations versus the dimension m of example 1. Here k is the number of inner iteration in (17), (18).

Iter.	$A_1 = D$				Jacobi method	$A_1 = A_L + D$				SOR
m/k	0	1	2	3		0	1	2	3	ω_{opt}
10	16	10	9	6	30	12	7	5	4	12
100	18	9	8	5	32	14	7	5	4	17
1000	17	9	8	5	34	14	7	5	4	18

Table 3. The behavior of the iteration parameter τ_n given by (18) for example 1. Here n and k are the numbers of outer and inner iterations in (17), respectively.

n/k	1	2	3
1	1.031939	0.994921	1.000760
2	0.998348	1.001490	0.998875
3	0.988243	0.999830	1.000464
4	0.989728	1.011261	0.999378
5	1.014757	1.004592	
6	1.024398		
7	1.053859		

Table 4. The number n of outer iteration with fixed number k of inner iteration, and CPU time CT for examples 2 and 3.

A_1		**Example 2**			**Example 3**		
		$k = 0$	$k = 1$	$k = 2$	$k = 0$	$k = 1$	$k = 2$
$\boldsymbol{D}$	n	39	17	13	196	89	56
	CT	1.5e–3	9.2e–4	8.6e–4	8.6e–4	6.9e–3	3.1e–3
$\boldsymbol{A_L + D}$	n	14	5	13	92	58	41
	CT	6.7e–3	4.9e–4	4.5e–4	3.3e–4	2.7e–3	2.3e–3
SOR	n		8			35	
	CT		1.3e–3			5.1e–3	

Table 5. The number n of outer iteration, the total number k of inner iteration when forcing terms η_n was chosen by formulas (32) and (33), and CPU time CT.

A_1			Example 2	Example 3
D	(33)	CT	1.61e–3	1.07e–2
		n (k)	4 (12)	5 (188)
	(32)	CT	1.72e–3	1.08e–2
		n (k)	5 (11)	5 (135)
$A_L + D$	(33)	CT	1.10e–3	7.53e–3
		n (k)	4 (5)	5 (94)
	(32)	CT	1.11e–3	7.42e–3
		n (k)	4 (4)	4 (92)

Table 6. A comparison of the calculated values u(1/2, 1/2) and an exact value (34) versus the dimension N of the example 4.

$N + 1$	$u(1/2, 1/2)$
4	0.070312
8	0.072783
16	0.073446
32	0.073615
Exact	0.073671

Table 7. The number n of outer iteration with fixed number k of inner iteration, and CPU time CT versus the dimension N of the example 4.

$N+1$		$A_1=D$			$A_1=A_L+D$			Tridiagonal			SOR
		$k=0$	$k=1$	$k=2$	$k=0$	$k=1$	$k=2$	$k=0$	$k=1$	$k=2$	
4	n	64	20	21	32	15	11	39	16	12	11
	CT	2.5e–3	1.3e–3	1.4e–3	1.3e–3	8.8e–4	8.1e–4	1.2e–3	8.3e–4	7.6e–4	1.3e–3
8	n	267	60	87	124	60	40	149	40	47	23
	CT	1.4e–2	9.6e–3	1.2e–3	6.7e–3	3.9e–3	4.2e–3	7.9e–3	2.8e–3	3.6e–3	6.1e–3
16	n	1010	159	328	476	214	137	546	99	183	46
	CT	5.9e–1	1.4e–1	3.0e–1	2.8e–1	1.7e–1	1.3e–1	2.9e–1	1.0e–1	1.8e–1	1.0e–1

Table 8. The number n of outer iteration with unfixed number k of inner iteration, the itration parameter τ_n, and CPU time CT versus the dimension N of the example 4.

$N+1$	n	$A_1=D$			$A_1=A_L+D$			Tridiagonal		
		k	τ_n	CT	k	τ_n	CT	k	τ_n	CT
	1	9	1.0304		5	1.0205		5	1.0267	
4	2	4	1.0087	3.58e–3	3	0.9845	2.42e–3	2	1.0207	2.22e–3
	3	18	1.0013		10	1.0006		8	1.0037	
	1	44	1.0235		23	1.0202		23	1.0213	
8	2	35	1.0108	3.37e–3	14	1.0118	1.67e–3	18	1.0102	1.77e–2
	3	70	1.0030		34	1.0032		35	1.0034	
	1	211	1.0122		106	1.0120		106	1.0122	
16	2	185	1.0072	6.07e–0	89	1.0075	3.07e–0	93	1.0072	3.10e–0
	3	195	1.0167		99	1.0158		98	1.0165	

4. CONCLUSIONS

Our method with inner iteration is quadratically convergent and therefore it can compete with other iterations such as SOR with an optimal relaxation parameter for a strictly diagonally dominant system. Moreover, our method is also applicable not only for the system with a strictly diagonal dominant matrix, but also for system, the matrix of which is not Hermitian and positive definite.

This work was partially sponsored by foundation for science and technology of Ministry of Education, Culture, and Science (Mongolia).

O. Chuluunbaatar acknowledges a financial support from the RFBR Grant No. 11-01-00523, and the theme 09-6-1060-2005/2013 "Mathematical support of experimental and theoretical studies conducted by JINR".

REFERENCES

1. R. Kress, "Numerical Analysis," Springer-Verlag, Berlin, 1998.
2. T. Zhanlav, "On the Iteration Method with Minimal Defect for Solving a System of Linear Algebraic Equations," Scientific Transaction, No. 8, 2001, pp. 59-64.
3. T. Zhanlav and I. V. Puzynin, "The Convergence of Iterations Based on a Continuous Analogy Newton's Method," Journal of Computational Mathematics and Mathematical Physics, Vol. 32, No. 6, 1992, pp. 729-737.
4. R. S. Dembo, S. C. Eisenstat and T. Steihaug, "Inexact Newton Methods," SIAM Journal on Numerical Analysis, Vol. 19, No. 2, 1982, pp. 400-408.
5. H. B. An, Z. Y. Mo and X. P. Liu, "A Choice of Forcing Terms in Inexact Newton Method," Journal of Computational and Applied Mathematics, Vol. 200, No. 1, 2007, pp. 47-60.
6. T. Zhanlav, O. Chuluunbaatar and G. Ankhbayar, "Relationship between the Inexact Newton Method and the Continuous Analogy of Newton Method," Revue D'Analyse Numerique et

de Theorie de L'Approximation, Vol. 40, No. 2, 2011, pp. 182-189.

7. T. Zhanlav and O. Chuluunbaatar, "The Local and Global Convergence of the Continuous Analogy of Newton's Method," Bulletin of PFUR, Series Mathematics, Information Sciences, Physics, No. 1, 2012, pp. 34-43.

8. J. W. Demmel, "Applied Numerical Linear Algebra," SIAM, Philadelphia, 1997, pp. 265-360.

9. W. Hackbusch, "Elliptic Differential Equations: Theory and Numerical Treatment," Springer Series in Computational Mathematics, Vol. 18, Springer, Berlin, 1992.

12

CHAPTER

Chebyshev Polynomials for Solving a Class of Singular Integral Equations

Samah M. Dardery[1], Mohamed M. Allan[2]

[1]*Department of Mathematics, Faculty of Science, Zagazig University, Zagazig, Egypt*

[2]*Department of Mathematics, Faculty of Science and Arts Al-Mithnab, Qassim University, Qassim, KSA*

ABSTRACT

This paper is devoted to study the approximate solution of singular integral equations by means of Chebyshev polynomials. Some examples are presented to illustrate the method.

Keywords: Singular Integral Equations; Cauchy Kernel; Chebyshev Polynomials; Weight Functions

1. INTRODUCTION

During the last three decades, the singular integral equation methods with applications to several basic fields of engineering mechanics, like elasticity, plasticity, aerodynamics and fracture mechanics have been studied and improved by several scientists (see [1] -[6]). Hence, it is interest to solve numerically this type of integral equations (see [7] [8]). Chebyshev polynomials are of great importance in many areas of mathematics particularly approximation theory (see [9] [10]).

In this paper we analyze the numerical solution of singular integral equations by using Chebyshev polynomials of first, second, third and fourth kind to obtain systems of linear algebraic equations, these systems are solved numerically. The methodology of the present work expected to be useful for solving singular integral equations of the first kind, involving partly singular and partly regular kernels. The singularity is assumed to be of the Cauchy type. The method is illustrated by considering some examples.

Singular integral equation of first kind, with a Cauchy type singular kernel, over a finite interval can be represented by

$$\int_{-1}^{1} \frac{k(t,x)\varphi(t)}{t-x} \mathrm{d}t + \int_{-1}^{1} L(t,x)\varphi(t)\mathrm{d}t = f(x), -1 < x < 1 \tag{1.1}$$

where $k(t,x), L(t,x)$ and $f(x)$ are given real-valued continuous functions belonging to the class Holder of continues functions and $k(t,t) \neq 0$. In Equation (1.1) the singular kernel is interpreted as Cauchy principle value. Integral equation of form (1.1) and other different forms have many applications (see [1] [2] [6] [11] [12]). The theory of this equation is well known and it is presented in [13] [14] . An approximate method for solving (1.1) using a polynomial approximation of degree n has been proposed in [7] .

It is well known that the analytical solutions of the simple singular integral equation

$$\int_{-1}^{1} \frac{\varphi(t)}{t-x} \mathrm{d}t = f(x), -1 < x < 1 \tag{1.2}$$

at $k(t,x) = 1$ and $L(t,x) = 0$, for the following four cases:

1) The solution is unbounded at both end-points $x = \pm 1,$

2) The solution is bounded at both end-points $x = \pm 1,$

3) The solution is bounded at end $x = -1,$ but unbounded at end $x = +1$ 4) The solution is unbounded at end $x = -1,$ but bounded at end $x = +1$ are given by [15] . In this paper the used approximate method for solving Equation (1.1) stems from recent work [10] wherein an approximate method has been developed to solve the simple Equation (1.2). The approximate method developed below appears to be quite appropriate for solving the most general type Equation (1.1). Some examples are presented to illustrate the method.

2. THE APPROXIMATE SOLUTION

In this section we present the method of the approximate solution of Equation (1.1) in four cases.

Let the unknown function φ in Equation (1.1) be approximated by the polynomial function

$$\varphi_n(x)=W^{(j)}(x)\sum_{i=0}^{n}c_i^{(j)}\Psi_i^{(j)}(x),(j=1,2,3,4) \tag{2.1}$$

where $c_i^{(j)},i=0,1,2,\cdots,n$ are unknown coefficients, to be determined, and in case (I): $\Psi_i^{(1)}(x)=T_i(x)$, in case (II): $\Psi_i^{(2)}(x)=U_i(x)$, in case (III): $\Psi_i^{(3)}(x)=V_i(x)$ and in case (VI): $\Psi_i^{(4)}(x)=W_i(x)$, where T_i, U_i, V_i and $W_i,i=0,1,\cdots,n,$ are The Chebyshev polynomials of the first, second, third and fourth kinds respectively can be defined by the recurrence relations [9] [16] .

$$\left.\begin{array}{l}T_0(x)=1,T_1(x)=x\\T_n(x)=2xT_{n-1}(x)-T_{n-2}(x),n\geq 2\end{array}\right\} \tag{2.2}$$

$$\left.\begin{array}{l}U_0(x)=1,U_1(x)=2x\\U_n(x)=2xU_{n-1}(x)-U_{n-2}(x),n\geq 2\end{array}\right\} \tag{2.3}$$

$$\left.\begin{array}{l}V_0(x)=1,V_1(x)=2x-1\\V_n(x)=2xV_{n-1}(x)-V_{n-2}(x),n\geq 2\end{array}\right\} \tag{2.4}$$

$$\left.\begin{array}{l}W_0(x)=1,W_1(x)=2x+1\\W_n(x)=2xW_{n-1}(x)-W_{n-2}(x),n\geq 2\end{array}\right\} \tag{2.5}$$

and $W^i, i = 0,1,\cdots,n,$ are the corresponding weight functions.

Substituting the approximate solution (2.1) for the unknown function into (1.1) yields

$$\sum_{i=0}^{n} c_i^{(j)} \left[\int_{-1}^{1} \frac{k(t,x) W^{(j)}(t) \Psi_i^{(j)}(t)}{t-x} \mathrm{d}t + \int_{-1}^{1} L(t,x) W^{(j)}(t) \Psi_i^{(j)}(t) \mathrm{d}t \right] = f(x), -1 < x < 1$$

(2.6)

In above Equation (2.6), we next use the following Chebyshev approximation to the kernels $k(t,x)$ and $L(t,x)$, given by (for fixed x, cf. [7])

$$k(t,x) \cong \sum_{p=0}^{m} k_p(x) t^p, L(t,x) \cong \sum_{q=0}^{s} L_q(x) t^q$$ (2.7)

with known expressions for $K_p(x)$ and $L_q(x)$. Then (2.6) gives

$$\sum_{i=0}^{n} c_i^{(j)} \alpha_i^{(j)}(x) = f(x), -1 < x < 1, (j = 1,2,3,4)$$

(2.8)

where

$$\alpha_i^{(j)}(x) = \sum_{p=0}^{m} k_p(x) u_{p,i}^{(j)}(x) + \sum_{q=0}^{s} L_q(x) \gamma_{q,i}^{(j)}$$ (2.9)

with

$$u_{p,i}^{(j)}(x)=\int_{-1}^{1}\frac{t^{p}W^{(j)}(t)\Psi_i^{(j)}(t)}{t-x}\mathrm{d}t,-1<x<1,(j=1,2,3,4)$$

(2.10)

and

$$\gamma_{q,i}^{(j)}=\int_{-1}^{1}t^{q}W^{(j)}(t)\Psi_i^{(j)}(t)\mathrm{d}t$$

(2.11)

Let $x_k^{(j)},j=1,2,3,4$, be the zeros of $U_n(x),T_{n+2}(x),W_{n+1}(x)$ and $V_{n+1}(x)$, respectively. Substituting the collocation points $x_k^{(j)},j=1,2,3,4$ into (2.8) we obtain the following systems of linear equations:

$$\sum_{i=0}^{n}c_i^{(j)}\alpha_i^{(j)}\left(x_k^{(j)}\right)=f\left(x_k^{(j)}\right),(k=1,2,\cdots,n+1),(j=1,2,3,4)$$

(2.12)

where

$$\alpha_i^{(j)}\left(x_k^{(j)}\right)=\sum_{p=0}^{m}k_p\left(x_k^{(j)}\right)u_{p,i}^{(j)}\left(x_k^{(j)}\right)+\sum_{q=0}^{s}L_q\left(x_k^{(j)}\right)\gamma_{q,i}^{(j)},(k=1,2,\cdots,n+1),(j=1,2,3,4)$$

(2.13)

Solving the system of Equation (2.12) for the unknown coefficients $c_i^{(j)},j=1,2,3,4$, and substituting the values of $c_i^{(j)}$ into (2.1) we obtain the approximate solutions of Equation (1.1).

3. NUMERICAL EXAMPLES

In this section, we consider some problems to illustrate the above method. All results were computed using FORTRAN code.

Example 1 Consider the following singular integral equation

$$\int_{-1}^{1}\frac{\left(x+t^2\right)\varphi(t)}{t-x}+\int_{-1}^{1}\left(x^2+t^3\right)\varphi(t)\,\mathrm{d}t=2x^4-2x^2-\frac{3}{8},-1<x<1$$

(3.1.1)

where

$k\left(x,t\right)=x+t^2, L\left(x,t\right)=x^2+t^3, f\left(t\right)=2x^4-2x^2-\frac{3}{8}$. So,

one gets

$$k_0\left(x\right)=x, k_1\left(x\right)=0, k_2\left(x\right)=1, k_p\left(x\right)=0, \left(p>2\right)$$

$$L_0\left(x\right)=x^2, L_1\left(x\right)=0, L_2\left(x\right)=0, L_3\left(x\right)=1, L_q\left(x\right)=0, \left(q>3\right)$$

Hence we find that relation (2.8) produces

$$\sum_{i=0}^{n}c_i^{(j)}\alpha_i^{(j)}\left(x\right)=2x^4-2x^2-\frac{3}{8},-1<x<1,\left(j=1,2,3,4\right)$$

Thus (2.9) gives

$$\alpha_i^{(j)}(x) = x u_{0,i}^{(j)}(x) + u_{2,i}^{(j)}(x) + x^2 \gamma_{0,i}^{(j)} + \gamma_{3,i}^{(j)}, (j = 1,2,3,4), (i = 0,1,2,\cdots)$$

Firstly, let us consider in detail the case (I), $j = 1$, for $n = 3$. This results in

$$u_{0,.i}^{(1)}(x) = \int_{-1}^{1} \frac{T_i(t)}{\sqrt{1-t^2} t - x} dt, u_{2,.i}^{(1)}(x) = \int_{-1}^{1} \frac{t^2 T_i(t)}{\sqrt{1-t^2} t - x} dt, -1 < t < 1, \tag{3.1.2}$$

$$\gamma_{0,i}^{(1)} = \int_{-1}^{1} \frac{T_i(t)}{\sqrt{1-t^2}} dt, \ \gamma_{3,i}^{(1)} = \int_{-1}^{1} \frac{t^3 T_i(t)}{\sqrt{1-t^2}} dt, \tag{3.1.3}$$

By applying the following relations

$$\int_{-1}^{1} \frac{T_i(t)}{\sqrt{1-t^2}(t-x)} dt = \pi U_{i-1}(x), \int_{-1}^{1} \frac{1}{\sqrt{1-t^2}(t-x)} dt = 0 \tag{3.1.4}$$

$$\int_{-1}^{1} \frac{T_i(t) T_j(t)}{\sqrt{1-t^2}} dt = \begin{cases} 0 & i \neq j \\ \pi & i = j = 0 \\ \pi/2 & i = j \neq 0 \end{cases} \tag{3.1.5}$$

It is easy to estimate the values $u_{0,i}^{(1)}, u_{2,i}^{(1)}, \gamma_{0,i}^{(1)}$ and $\gamma_{3,i}^{(1)}$.

From (2.2) and (3.1.2)-(3.1.5) we get

$$\alpha_i^{(1)}(x)=\begin{cases}\pi\left(x^2+x\right); & i=0\\ \pi\left(x^2+x+\dfrac{7}{8}\right); & i=1\\ \pi\left(2x^3+2x^2\right); & i=2\\ \pi\left(4x^4+4x^3-x^2-x+\dfrac{1}{8}\right); & i=3\end{cases}$$

(3.1.6)

By choosing the collocation points $x_k=\cos\left(\dfrac{(2k-1)\pi}{2(n+2)}\right),(k=1,2,3,4)$, for $n=3,$ we obtain the following system of linear equations:

$$\sum_{i=0}^{3}c_i^{(1)}\alpha_i^{(1)}(x_k)=f(x_k),k=1,2,3,4$$

By solving this system for the unknown coefficients $c_i^{(1)},i=0,1,2,3$ that produces

$$\left.\begin{aligned}c_0^{(1)}&=0.3183098,c_1^{(1)}=-0.1591549\\ c_2^{(1)}&=-0.3183098,c_3^{(1)}=0.1591549\end{aligned}\right\}$$

(3.1.7)

From (3.1.7) we obtain the approximate solution of Equation (3.1.1) in the form

$$\varphi_n(x) \cong \frac{2}{\pi\sqrt{1-x^2}}\left(x^3 - x^2 - x + 1\right) \tag{3.1.8}$$

Which coincides with the exact solution. The error of approximate solution (3.1.8) of Equation (3.1.1) at $n = 20$ is given by Table 1.

Secondly, let us consider in detail the case (II), $j = 2$, for $n = 3$. This results in

$$u_{0,i}^{(2)}(x) = \int_{-1}^{1} \frac{\sqrt{1-t^2}U_i(t)}{t-x}\mathrm{d}t, u_{2,i}^{(2)}(x) = \int_{-1}^{1} \frac{t^2\sqrt{1-t^2}U_i(t)}{t-x}\mathrm{d}t, -1 < t < 1, \tag{3.1.9}$$

$$\gamma_{0,i}^{(2)} = \int_{-1}^{1} \sqrt{1-t^2}U_i(t)\mathrm{d}t,\ \gamma_{3,i}^{(2)} = \int_{-1}^{1} \sqrt{1-t^2}t^3U_i(t)\mathrm{d}t, \tag{3.1.10}$$

By applying the following relations

$$\int_{-1}^{1} \frac{\sqrt{1-t^2}U_i(t)}{t-x}\mathrm{d}t = -\pi T_{i+1}(x) \tag{3.1.11}$$

$$\int_{-1}^{1} \sqrt{1-t^2}U_i(t)U_j(t)\mathrm{d}t = \begin{cases} 0 & i \neq j \\ \dfrac{\pi}{2} & i = j \end{cases} \tag{3.1.12}$$

It is easy to estimate the values $u_{0,i}^{(2)}, u_{2,i}^{(2)}, \gamma_{0,i}^{(2)}$ and $\gamma_{3,i}^{(2)}$.

From the relations (2.3) and (3.1.9)-(3.1.12) we get

$$\alpha_i^{(2)}(x)=\begin{cases}-\frac{\pi}{2}\left(2x^3+x^2-x\right) & i=0\\ -\pi\left(2x^4+2x^3-x^2-x-\frac{3}{8}\right) & i=1\\ -\pi\left(4x^5+4x^4-3x^3-3x^2+\frac{1}{4}x\right) & i=2\\ -\pi\left(8x^6+8x^5-8x^4-8x^3+x^2+x-\frac{1}{16}\right) & i=3\end{cases}$$

(3.1.13)

By choosing the collocation points $x_k^{(2)}=\cos\left(\frac{(2k-1)\pi}{2(n+2)}\right),(k=1,2,3,4)$, for $n=3,$ we obtain the following system of linear equations :

$$\sum_{i=0}^{3} c_i^{(2)}\alpha_i^{(2)}\left(x_k^{(2)}\right)=f\left(x_k^{(2)}\right),k=1,2,3,4$$

By solving this system for the unknown coefficients $c_i^{(2)},i=0,1,2,3$ that produces

$$\left.\begin{array}{l}c_0^{(2)}=0.6366197,c_1^{(2)}=-0.3183099\\ c_2^{(2)}=2.279989\times10^{-8},c_3^{(2)}=-7.819254\times10^{-9}\end{array}\right\}$$

(3.1.14)

From (3.1.14) we obtain the approximate solution of Equation (3.1.1) in the form

$$\varphi_n(x) \cong \frac{2\sqrt{1-x^2}}{\pi}(1-x) \tag{3.1.15}$$

Which coincides with the exact solution. The error of approximate solution (3.1.15) of Equation (3.1.1) at $n=20$ is given by Table 1.

Thirdly, let us consider in detail the case (III), $j=3$, for $n=3$. This results in

$$u_{0,i}^{(3)}(x) = \int_{-1}^{1}\sqrt{\frac{1+t}{1-t}}\frac{V_i(t)}{t-x}\mathrm{d}t, u_{2,i}^{(3)}(x) = \int_{-1}^{1}\sqrt{\frac{1+t}{1-t}}\frac{t^2V_i(t)}{t-x}\mathrm{d}t, -1<t<1, \tag{3.1.16}$$

$$\gamma_{0,i}^{(3)} = \int_{-1}^{1}\sqrt{\frac{1+t}{1-t}}V_i(t)\mathrm{d}t,\ \gamma_{3,i}^{(3)} = \int_{-1}^{1}\sqrt{\frac{1+t}{1-t}}t^3V_i(t)\mathrm{d}t, \tag{3.1.17}$$

By applying the following relations

$$\int_{-1}^{1}\sqrt{\frac{1+t}{1-t}}V_i(t)V_j(t)\mathrm{d}t = \begin{cases} 0 & i \neq j \\ \pi & i = j \end{cases} \tag{3.1.18}$$

$$\int_{-1}^{1}\sqrt{\frac{1+t}{1-t}}\frac{V_i(t)}{t-x}\mathrm{d}t = \pi W_i(x) \tag{3.1.19}$$

It is easy to estimate the values $u_{0,i}^{(3)}, u_{2,i}^{(3)}, \gamma_{0,i}^{(3)}$ and $\gamma_{3,i}^{(3)}$.

Table 1. Illustrates errors of approximate solutions of Equation (3.1.1) in Cases (I)-(IV) at n = 20.

x	error (j = 1)	error (j = 2)	error (j = 3)	error (j = 4)
−9.500000E-01	0.000000E+00	0.000000E+00	5.960464E-08	5.960464E-08
−9.000000E-01	0.000000E+00	0.000000E+00	1.192093E-07	1.192093E-07
−7.0000000E-01	0.000000E+00	0.000000E+00	1.192093E-07	1.192093E-07
−5.000000E-01	5.960464-08	5.960464E-08	1.788139E-07	1.788139E-07
−3.000000E-01	0.000000E+00	5.960464E-08	1.788139E-07	1.788139E-07
−1.000000E-01	5.960464E-08	5.960464E-08	1.192093E-07	1.192093E-07
0.000000E+00	5.960464E-08	5.960464E-08	1.192093E-07	1.192093E-07
1.000000E-01	1.192093E-07	5.960464E-08	5.960464E-08	5.960464E-08
3.000000E-01	8.940697E-08	8.940697E-08	8.940697E-08	8.940697E-08
5.000000E-01	8.940697E-08	8.940697E-08	5.960464E-08	5.960464E-08
7.000000E-01	1.043081E-07	7.450581E-08	1.490116E-08	1.490116E-08
9.000000E-01	9.313226E-08	5.029142E-08	4.656613E-08	1.303852E-08
9.500000E-01	5.774200E-08	3.632158E-08	6.705523E-08	4.656613E-09

From the relations (2.4) and (3.1.16)-(3.1.19) we get

$$\alpha_i^{(3)}(t)=\begin{cases}\pi\left(2x^2+2x+\dfrac{7}{8}\right) & i=0\\ \pi\left(2x^3+3x^2+x+\dfrac{7}{8}\right) & i=1\\ \pi\left(4x^4+6x^3+x^2-x+\dfrac{1}{8}\right) & i=2\\ \pi\left(8x^5+12x^4-5x^2+x+\dfrac{1}{8}\right) & i=3\end{cases} \tag{3.1.20}$$

By choosing the collocation points $x_k^{(3)} = \cos\left(\frac{2k\pi}{(2n+3)}\right), (k = 1,2,3,4)$, for $n = 3,$ we obtain the following system of linear equations :

$$\sum_{i=0}^{3} c_i^{(3)} \alpha_i^{(3)} \left(x_k^{(3)}\right) = f\left(x_k^{(3)}\right), k = 1,2,3,4$$

By solving this system for the unknown coefficients $c_i^{(3)}, i = 0,1,2,3$ that produces

$$\left. \begin{array}{l} c_0^{(3)} = 0.3183098, c_1^{(3)} = -0.4774647, \\ c_2^{(3)} = 0.1591549, c_3^{(3)} = 1.330901 \times 10^{-8} \end{array} \right\} \tag{3.1.21}$$

From (3.1.21) we obtain the approximate solution of Equation (3.1.1) in the form of

$$\varphi_n(x) \cong \frac{2}{\pi} \sqrt{\frac{1+x}{1-x}} \left(x^2 - 2x + 1\right) \tag{3.1.22}$$

Which coincides with the exact solution. The error of approximate solution (3.1.22) of Equation (3.1.1) at $n = 20$ is given by Table 1.

Fourthly, In case (IV), $j = 4$, for $n = 3$. This results in

$$u_{0,.i}^{(4)}(x)=\int_{-1}^{1}\sqrt{\frac{1-t}{1+t}}\frac{W_i(t)}{t-x}\mathrm{d}t, u_{2,.i}^{(4)}(x)=\int_{-1}^{1}\sqrt{\frac{1-t}{1+t}}\frac{t^2W_i(t)}{t-x}\mathrm{d}t, -1<t<1,$$

(3.1.23)

$$\gamma_{0,i}^{(4)}=\int_{-1}^{1}\sqrt{\frac{1-t}{1+t}}W_i(t)\mathrm{d}t, \gamma_{3,i}^{(4)}=\int_{-1}^{1}\sqrt{\frac{1-t}{1+t}}t^3W_i(t)\mathrm{d}t,$$

(3.1.24)

By applying the relations

$$\int_{-1}^{1}\sqrt{\frac{1-t}{1+t}}W_i(t)W_j(t)\mathrm{d}t=\begin{cases}0 & i\neq j\\ \pi & i=j\end{cases}$$

(3.1.25)

$$\int_{-1}^{1}\sqrt{\frac{1-t}{1+t}}\frac{W_i(t)}{t-x}\mathrm{d}t=-\pi V_i(x)$$

(3.1.26)

It is easy to estimate the values $u_{0,i}^{(4)}, u_{2,i}^{(4)}, \gamma_{0,i}^{(4)}$ and $\gamma_{3,i}^{(4)}$.

From the relations (2.5) and (3.1.23)-(3.1.26) we get

$$\alpha_i^{(4)}(x)=\begin{cases}\dfrac{-7\pi}{8}; & i=0\\ -\pi\left(2x^3+x^2-x-\dfrac{7}{8}\right); & i=1\\ -\pi\left(4x^4+2x^3-3x^2-x+\dfrac{1}{8}\right); & i=2\\ -\pi\left(8x^5+4x^4-8x^3-3x^2+x-\dfrac{1}{8}\right); & i=3\end{cases}$$

(3.1.27)

By choosing the collocation points $x_k^{(4)}=\cos\left(\dfrac{(2k-1)\pi}{(2n+3)}\right),(k=1,2,3,4)$, for $n=3,$ we obtain the following system of linear equations :

$$\sum_{i=0}^{3}c_i^{(4)}\alpha_i^{(4)}\left(x_k^{(4)}\right)=f\left(x_k^{(4)}\right),k=1,2,3,4$$

By solving this system for the unknown coefficients $c_i^{(4)},i=0,1,2,3$ that produces

$$\left.\begin{aligned}c_0^{(4)}&=0.3183098,c_1^{(4)}=0.1591549\\ c_2^{(4)}&=-0.1591549,c_3^{(4)}=2.358931\times10^{-8}\end{aligned}\right\}\quad(3.1.28)$$

From (3.1.28) we obtain the approximate solution of Equation (3.1.1) in the form of

$$\varphi_n(x) \cong \frac{-2}{\pi}\sqrt{\frac{1-x}{1+x}}\left(x^2-1\right) \tag{3.1.29}$$

which coincides with the exact solution. The error of approximate solution (3.2.29) of Equation (3.2.1) at $n=20$ is given by Table 1.

Example 2. Consider the following singular integral equation

$$\int_{-1}^{1}\frac{\varphi(t)}{t-x}\,dt + \int_{-1}^{1}\left(x^3+xt^2\right)\varphi(t)\,dt = x^3+x \tag{3.2.1}$$

which corresponds with $k(t,x)=1$ and $L(t,x)=x^3+xt^2$. So one gets

$$k_0(x)=1, k_p(x)=0, (p>0)$$

$$L_0(x)=x^3, L_1(x)=0, L_2(x)=x, L_q(x)=0(q>2)$$

Hence we find that relation (2.8) produces

$$\sum_{i=0}^{n} c_i^{(j)}\alpha_i^{(j)}(x) = x^3+x, -1<x<1, (j=1,2,3,4)$$

Thus (2.9) gives

$$\alpha_i^{(j)}(x) = u_{0,i}^{(j)}(x) + x^3\gamma_{0,i}^{(j)} + x\gamma_{2,i}^{(j)}, (j=1,2,3,4), (i=0,1,2,\cdots)$$

Firstly, let us consider in detail the case (I), $j=1$, for $n=3$. This results in

$$\gamma_{2,i}^{(1)} = \int_{-1}^{1} \frac{t^2 T_i(t)}{\sqrt{1-t^2}} \mathrm{d}t, \tag{3.2.2}$$

From the relations (3.1.2)-(3.1.5) and (3.2.2) we obtain

$$\alpha_i^{(1)}(x) = \begin{cases} \pi\left(x^3 + x/2\right) & i=0 \\ \pi & i=1 \\ 9\pi x/4 & i=2 \\ \pi\left(4x^2-1\right) & i=3 \end{cases} \tag{3.2.3}$$

By choosing the collocation points $x_k = \cos\left(\frac{(2k-1)\pi}{2(n+2)}\right), (k=1,2,3,4)$, for $n=3$, we obtain the following system of linear equations :

$$\sum_{i=0}^{3} c_i^{(1)} \alpha_i^{(1)}(x_k) = f(x_k), k=1,2,3,4$$

By solving this system for the unknown coefficients $c_i^{(1)}, i=0,1,2,3$ that produces

$$\left.\begin{matrix} c_0^{(1)} = 0.3183098, c_1^{(1)} = 1.090772\times 10^{-8} \\ c_2^{(1)} = 0.07073557, c_3^{(1)} = 1.830649\times 10^{-8} \end{matrix}\right\} \tag{3.2.4}$$

From (3.2.4) we obtain the approximate solution of Equation (3.2.1) in the form of

$$\varphi_n(x) \cong \frac{1}{9\pi\sqrt{1-x^2}}\left(7+4x^2\right) \tag{3.2.5}$$

which coincides with the exact solution. The error of approximate solution (3.2.5) of equation (3.2.1) at $n = 20$ is given by Table 2.

Secondly, let us consider in detail the case (II), $j = 2$, for $n = 3$. This results in

$$\gamma_{2,i}^{(2)} = \int_{-1}^{1} \sqrt{1-t^2}\,t^2 U_i(t)\,dt, \tag{3.2.6}$$

By applying the relations (3.1.9)-(3.1.12) and (3.2.6) we get

$$\alpha_i^{(2)}(x) = \begin{cases} \frac{\pi}{2}\left(x^3 - \frac{7x}{4}\right) & i=0 \\ -\pi\left(2x^2-1\right) & i=1 \\ -\pi\left(4x^3 - \frac{25x}{8}\right) & i=2 \\ -\pi\left(8x^4-8x^2+1\right) & i=3 \end{cases} \tag{3.2.7}$$

By choosing the collocation points $x_k^{(2)} = \cos\left(\frac{(2k-1)\pi}{2(n+2)}\right), (k=1,2,3,4)$, for $n = 3,$ we obtain the following system of linear equations:

Table 2. Illustrates errors of approximate solutions of Equation (3.2.1) in Case (I), Case (II) and Case (IV) respectively at n = 20.

x	Error ($j=1$)	Error ($j-2$)	Error ($j=3$)
−9.500000E-01	0.000000E+00	0.000000E+00	0.000000E+00
−9.000000E-01	5.960464E-08	5.960464E-08	0.000000E+00
−7.000000E-01	8.940697E-08	1.192093E-07	5.960464E-08
−5.000000E-01	8.940697E-08	1.192093E-07	1.192093E-07
−3.000000E-01	8.940697E-08	1.788139E-07	1.192093E-07
−1.000000E-01	1.192093E-07	1.788139E-07	1.788139E-07
0.000000E+00	1.043081E-07	1.788139E-07	1.788139E-07
1.000000E-01	1.192093E-07	1.788139E-07	1.192093E-07
3.000000E-01	8. 940697E-08	1.788139E-07	5.960464E-08
5.000000E-01	8.940697E-08	1.192093E-07	1.192093E-07
7.000000E-01	8.940697E-08	1.192093E-07	0.00000E+00
9.000000E-01	5.9604641E-08	5.960464E-08	5.960464E-08
9.500000E-01	0.000000E+00	0.000000E+00	0.000000E+00

$$\sum_{i=0}^{3} c_i^{(2)} \alpha_i^{(2)}\left(x_k^{(2)}\right) = f\left(x_k^{(2)}\right), k = 1,2,3,4$$

By solving this system for the unknown coefficients $c_i^{(2)}, i = 0,1,2,3$ that produces

$$\left.\begin{aligned} c_0^{(2)} &= -1.170559, c_1^{(2)} = -1.331665\times10^{-9} \\ c_2^{(2)} &= -0.2258973, c_3^{(2)} = -1.644008\times10^{-8} \end{aligned}\right\} \tag{3.2.8}$$

From (3.2.8) we obtain the approximate solution of Equation (3.2.1) in the form of

$$\varphi_n(x) \cong \frac{-\sqrt{1-x^2}}{31\pi}\left[92+88x^2\right] \tag{3.2.9}$$

which coincides with the exact solution. The error of approximate solution (3.2.9) of Equation (3.2.1) at $n = 20$ is given by Table 2.

Thirdly, In case (IV), $j = 4$, for $n = 3$. This results in

$$\gamma_{2,i}^{(4)} = \int_{-1}^{1} \sqrt{\frac{1-t}{1+t}} t^2 W_i(t) \mathrm{d}t, \tag{3.2.10}$$

By applying the relations (3.1.23)-(3.1.26) and (3.2.10) we get

$$\alpha_i^{(4)}(x) = \begin{cases} \pi\left(x^3 + \frac{x}{2} - 1\right) & i = 0 \\ -\pi\left(\frac{9}{4}x - 1\right) & i = 1 \\ -\pi\left(4x^2 - \frac{9x}{4} - 1\right) & i = 2 \\ -\pi\left(8x^3 - 4x^2 - 4x + 1\right) & i = 3 \end{cases} \tag{3.2.11}$$

By choosing the collocation points $x_k^{(4)} = \cos\left(\frac{(2k-1)\pi}{(2n+3)}\right), (k = 1,2,3,4)$, for $n = 3$, we obtain the following system of linear equations:

$$\sum_{i=0}^{3} c_i^{(4)} \alpha_i^{(4)}\left(x_k^{(4)}\right) = f\left(x_k^{(4)}\right), k = 1,2,3,4$$

By solving this system for the unknown coefficients $c_i^{(4)}, i = 0,1,2,3$ that produces

$$c_0^{(4)} = c_1^{(4)} = -0.5852794, c_2^{(4)} = c_3^{(4)} = -0.1129487 \quad (3.2.12)$$

From (3.2.12) we obtain the approximate solution of Equation (3.2.1) in the form of

$$\varphi_n(x) \cong \frac{-1}{31\pi}\sqrt{\frac{1-x}{1+x}}(1+x)\left(92+88x^2\right) \quad (3.2.13)$$

which coincides with the exact solution. The error of approximate solution (3.2.13) of Equation (3.2.1) at $n = 20$ is given by Table 2.

Similarly, doing the same operations as we did for Case (I), Case (II) and Case (IV), one can solve for Case (III) .

Example 3. Consider the following singular integral equation

$$\int_{-1}^{1} \frac{\varphi(t)}{t-x} dt + \int_{-1}^{1} \left(x^2 + t^2\right)\varphi(t) dt = \frac{-3}{2}x^2 + 2x \quad , (3.3.1)$$

which corresponds with $k(t,x) = 1$ and $L(t,x) = x^2 + t^2$. So, one gets

$$k_0(x) = 1, k_p(x) = 0, (p > 0)$$

$$L_0(x)=x^2, L_1(x)=0, L_2(x)=1, L_q(x)=0(q>2)$$

Hence the relation (2.8) produces

$$\sum_{i=0}^{n} c_i^{(j)} \alpha_i^{(j)}(x) = \frac{-3}{2}x^2 + 2x, -1 < x < 1, j = 1,2,3,4$$

(3.3.2)

where (2.9) gives

$$\alpha_i^{(j)}(x) = u_{0,i}^{(j)}(x) + x^2\gamma_{0,i}^{(j)} + \gamma_{2,i}^{(j)}, (j=1,2,3,4), (i=0,1,2,\cdots)$$

Firstly, let us consider in detail the case (II), $j=2$, for $n=3$. From (3.1.9)-(3.1.12) and (3.2.6) we get

$$\alpha_i^{(2)}(x) = \begin{cases} \frac{\pi}{8}\left(4x^2-8x+1\right) & i=0 \\ -\pi\left(2x^2-1\right) & i=1 \\ \frac{-\pi}{8}\left(32x^3-24x-1\right) & i=2 \\ -\pi\left(8x^4-8x^2+1\right) & i=3 \end{cases}$$

(3.3.3)

By solving the system (3.3.2), at the collocation points $x_k^{(2)} = \cos\left(\frac{(2k-1)\pi}{2(n+2)}\right), (k=1,2,3,4)$, for the unknown coefficients $c_i^{(2)}, i=0,1,2,3$ we obtain

$$\left.\begin{aligned} c_0^{(2)} &= -0.6366197, c_1^{(2)} = 0.07957754 \\ c_2^{(2)} &= 1.746461\times 10^{-8}, c_3^{(2)} = 1.827517\times 10^{-8} \end{aligned}\right\} \quad (3.3.4)$$

So the approximate solution of Equation (3.3.1) is given by

$$\varphi_n(x) \cong \frac{-\sqrt{1-x^2}}{2\pi}(4-x) \quad , (3.3.5)$$

which coincides with the exact solution, the error of the approximate solution (3.3.5) of Equation (3.3.1) at $n=20$ is given by Table 3.

Secondly, in case (III), $j=3$, for $n=3$. This results in

$$\gamma_{2,i}^{(3)} = \int_{-1}^{1} \sqrt{\frac{1+t}{1-t}} t^2 V_i(t)\,\mathrm{d}t, \quad (3.3.6)$$

From (3.1.16)-(3.1.19) and (3.3.6) we get

$$\alpha_i^{(3)}(x) = \begin{cases} \pi\left(x^2+\dfrac{3}{2}\right) & i=0 \\ \pi\left(2x+\dfrac{5}{4}\right) & i=1 \\ \pi\left(4x^2+2x-\dfrac{3}{4}\right) & i=2 \\ \pi\left(8x^3+4x^2-4x+1\right) & i=3 \end{cases} \quad (3.3.7)$$

By solving the system (3.3.2), at the collocation points $x_k^{(3)} = \cos\left(\frac{2k\pi}{(2n+3)}\right), (k = 1,2,3,4)$, for the unknown coefficients $c_i^{(3)}, i = 0,1,2,3$ we obtain

$$\left.\begin{array}{l} c_0^{(3)} = -0.3183099, c_1^{(3)} = 0.3580987, \\ c_2^{(3)} = -0.03978872, c_3^{(3)} = -8.155105 \times 10^{-9} \end{array}\right\} \quad (3.3.8)$$

Hence, the approximate solution of Equation (3.3.1) is given by

$$\varphi_n(x) \cong \frac{-1}{2\pi}\sqrt{\frac{1+x}{1-x}}\left(x^2 - 5x + 4\right) \quad (3.3.9)$$

which coincides with the exact solution, the error of the approximate solution (3.3.9) of Equation (3.3.1) at $n = 20$ is given by Table 3.

Similarly, doing the same operations as we did for Case (II) and Case (III), one can solve for Case (I) and Case (IV).

Table 3. Illustrates errors of approximate solutions of Equation (3.3.1) in Case (II) and Case (III) at n = 20.

x	Error (j = 2)	Error (3)
−9.500000E-01	2.980232E-08	2.980232E-08
−9.000000E-01	2.980232E-08	5.960464E-08
−7.000000E-01	0.000000E+00	5.960464E-08
−5.000000E-01	0.000000E+00	1.192093E-07
−3.000000E-01	0.000000E+00	1.192093E-07
−1.000000E-01	5.960464E-08	1.192093E-07
0.000000E+00	5.960464E-08	1.192093E-07
1.000000E-01	5.960464E-08	1.192093E-07
3.000000E-01	1.192093E-07	1.192093E-07
5.000000E-01	1.192093E-07	8.940697E-08
7.000000E-01	1.192093E-07	0.000000E+00
9.000000E-01	8.940697E-08	1.788139E-07
9.500000E-01	5.960464E-08	3.278255E-07

4. CONCLUSION

Numerical results (Tables 1-3) show that the errors of approximate solutions of Examples 1-3 in different Cases with small value of n are very small. These show that the methods developed are very accurate and in fact for a linear function give the exact solution.

REFERENCES

1. Chakrabarti, A. (1989) Solution of Two Singular Integral Equations Arising in Water Wave Problems. ZAMM, 69, 457-459.

2. Ladopoulous, E.G. (2000) Singular Integral Equations Linear and Non-Linear Theory and Its Applications in Science and Engineering. Springer, Berlin.

3. Ladopoulous, E.G. (1987) On the Solution of the Two-Dimensional Problem of a Plane Crack of Arbitrary Shape in an Anisotropic Material. Engineering Fracture Mechanics, 28, 187-195.

4. Zabreyko, P.P. (1975) Integral Equations—A Reference Text. Noordhoff, Leyden.

5. Prossdorf, S. (1977) On Approximate Methods for the Solution of One-Dimensional Singular Integral Equations. Applicable Analysis, 7, 259-270.

6. Zisis, V.A. and Ladopoulos, E.G. (1989) Singular Integral Approximations in Hilbert Spaces for Elastic Stress Analysis in a Circular Ring with Curvilinear Cracks. Indus. Math., 39, 113-134.

7. Chakrabarti, A. and Berghe, V.G. (2004) Approximate Solution of Singular Integral Equations. Applied Mathematics Letters, 17, 553-559.

8. Abdou, M.A. and Naser, A.A. (2003) On the Numerical Treatment of the Singular Integral Equation of the Second Kind. Applied Mathematics and Computation, 146, 373-380.

9. Abdulkawi, M., Eshkuvatov, Z.K. and Nik Long, N.M.A. (2009) A Note on the Numerical Solution of Singular Integral Equations of Cauchy Type. International Journal of Applied Mathematics and Computer Science, 5, 90-93.

10. Eshkuvatov, Z.K., Nik Long, N.M.A. and Abdulkawi, M. (2009) Approximate Solution of Singular Integral Equations of the First Kind with Cauchy Kernel. Applied Mathematics Letters, 22, 651-657.

11. Gakhov, F.D. (1966) Boundary Value Problems. Addison-Wesley, Boston.

12. Martin, P.A. and Rizzo, F.J. (1989) On Boundary Integral Equations for Crack Problems. Proceedings of the Royal Society A, 421, 341-345.

13. Sheshko, M. (2003) Singular Integral Equations with Cauchy and Hilbert Kernels and Their Approximated Solutions. The Learned Society of the Catholic University of Lublin, Lublin. (in Russian)

14. Muskhelishvili, N.I. (1977) Singular Integral Equations. Noordhoff International Publishing, Leyden.

15. Lifanov, I.K. (1996) Singular Integral Equation and Discrete Vortices. VSP, Leiden.

16. Kyth, K.P. and Schaferkotter, R.M. (2005) Handbook of Computational Methods for Integration. Chapman & Hall/ CRC Press, London.

13

CHAPTER

On an Operator Preserving Inequalities between Polynomials

Nisar Ahmad Rather, Mushtaq Ahmad Shah, Mohd. Ibrahim Mir

Post-Graduate Department of Mathematics, Kashmir University, Srinagar, India

ABSTRACT

Let P_n be the class of polynomials $P(z)$ of degree n and B_n a family of operators that map P_n into itself. For $B \in B_n$, we investigate the dependence of

$$\left| B[P(Rz)] - \alpha B[P(rz)] + \beta\left\{\left(\frac{R+1}{r+1}\right)^n - |\alpha|\right\} B[P(rz)] \right|$$

on the maximum modulus of $P(z)$ on $|z|=1$ for arbitrary real or complex numbers α, β with $|\alpha| \le 1$, $|\beta| \le 1$ and $R > r \ge 1$, and present certain sharp operator preserving inequa-lities between polynomials.

Keywords: Component Polynomials; B-Operator; Complex Domain

1. INTRODUCTION TO THE STATEMENT OF RESULTS

Let $P_n(z)$ denote the space of all complex polynomials $P(z)=\sum_{j=0}^{n} a_j z^j$ of degree n. If $P \in P_n$, then concerning the estimate of the maximum of $|P'(z)|$ on the unit circle $|z|=1$ and the estimate of the maximum of $|P(z)|$ on a larger circle $|z|=R>1$, we have

$$\max_{|z|=1}|P'(z)| \le n \max_{|z|=1}|P(z)| \tag{1}$$

and

$$\max_{|z|=R>1}|P(z)| \le R^n \max_{|z|=1}|P(z)|. \tag{2}$$

Inequality (1) is an immediate consequence of S. Bernstein's theorem (see [1-3]) on the derivative of a trigonometric polynomial. Inequality (2) is a simple deduction from the maximum modulus principle (see [4, p. 346] or [5, p. 158]). If we restrict ourselves to the class of polynomials $P \in P_n$ having no zero in $|z|<1$, then (1) and (2) can be replaced by

$$\max_{|z|=1}|P'(z)| \le \frac{n}{2}\max_{|z|=1}|P(z)| \tag{3}$$

and

$$\max_{|z|=R>1}|P(z)| \le \frac{R^n+1}{2}\max_{|z|=1}|P(z)|. \tag{4}$$

Inequality (3) was conjectured by Erdös and later verified by Lax [6]. Ankeny and Rivlin [7] used Inequality (3) to prove Inequality (4).

As a compact generalization of Inequalities (1) and (2), Aziz and Rather [8] have shown that if $P \in P_n$, then for every real or complex number α with $|\alpha| \le 1$, $R > 1$ and $|z| = 1$,

$$|P(Rz) - \alpha P(z)| \le |R^n - \alpha||z|^n \max_{|z|=1}|P(z)|. \tag{5}$$

The result is sharp.

As a corresponding compact generalization of Inequalities (3) and (4), they [8] have also shown that if $P \in P_n$, and $P(z) \ne 0$ for $|z| < 1$, then for every real or complex number α with $|\alpha| \le 1$, $R \ge 1$,

$$|P(Rz) - \alpha P(z)| \le \frac{1}{2}\left\{|R^n - \alpha||z|^n + |1-\alpha|\right\}\max_{|z|=1}|P(z)| \tag{6}$$

for $|z| \geq 1$. The result is sharp and equality in (6) holds for $P(z) = az^n + b$, $|a| = |b| = 1$.

Consider an operator B which carries a polynomial $P \in P_n$ into

$$B[P(z)] = \lambda_0 P(z) + \lambda_1 \left(\frac{nz}{2}\right) \frac{P'(z)}{1!} + \lambda_2 \left(\frac{nz}{2}\right)^2 \frac{P''(z)}{2!}, \tag{7}$$

where λ_0, λ_1 and λ_2 are such that all the zeros of

$$u(z) = \lambda_0 + \lambda_1 C(n,1) z + \lambda_2 C(n,2) z^2 \tag{8}$$

lie in the half plane

$$|z| \leq |z - n/2|. \tag{9}$$

As a generalization of the Inequalities (1) and (2), Q.I. Rahman [9] proved that if $P \in P_n$, then for $|z| \geq 1$,

$$|B[P(z)]| \leq |B[z^n]| \max_{|z|=1} |P(z)|. \tag{10}$$

and if $P(z) \neq 0$ for $|z| < 1$, then for $|z| \geq 1$,

$$|B[P(z)]| \leq \frac{1}{2}\{|B[z^n]| + |\lambda_0|\} \max_{|z|=1} |P(z)|. \tag{11}$$

(see [9], Inequality (5.2) and (5.3)).

In this paper, we consider a problem of investigating the dependence of

$$\left|B[P(Rz)]-\alpha B[P(rz)]+\beta\left\{\left(\frac{R+1}{r+1}\right)^n-|\alpha|\right\}B[P(rz)]\right|$$

on the maximum modulus of $P(z)$ on $|z|=1$ for arbitrary real or complex numbers α, β with $|\alpha|\leq 1$, $|\beta|\leq 1$ and $R>r\geq 1$, and develop a unified method for arriving at these results. In this direction we first present the following interesting result which is compact generalization of the Inequalities (1), (2), (5) and (10).

Theorem 1. If $P\in P_n$, then for arbitrary real or complex numbers α and β with $|\alpha|\leq 1$, $R>r\geq 1$ $|\beta|\leq 1$ and $|z|\geq 1$,

$$\begin{aligned}&\left|B[P(Rz)]+\phi(R,r,\alpha,\beta)B[P(rz)]\right|\\&\leq\left|R^n+\phi(R,r,\alpha,\beta)r^n\right|\left|B[z^n]\right|\max_{|z|=1}|P(z)|\end{aligned} \tag{12}$$

where

$$\phi(R,r,\alpha,\beta)=\beta\left\{\left(\frac{R+1}{r+1}\right)^n-|\alpha|\right\}-\alpha.$$

The result is best possible and equality in (12) holds for $P(z)=\lambda z^n;\ \lambda\neq 0.$

Remark 1. For $\beta=0,$ from Inequality (12), we have for $P\in P_n$, $|\alpha|\leq 1$, $|z|\geq 1$ and $R>r\geq 1,$

$$\left|B[P(Rz)]-\alpha B[P(rz)]\right|$$
$$\leq \left|R^n-\alpha r^n\right|\left|B[z^n]\right|\max_{|z|=1}|P(z)|. \tag{13}$$

Remark 2. For $\alpha=0$ and $\beta=0$, Inequality (12) reduces to

$$\left|B[P(Rz)]\right| \leq R^n\left|B[z^n]\right|\max_{|z|=1}|P(z)|$$
$$=\left|B[R^n z^n]\right|\max_{|z|=1}|P(z)|. \tag{14}$$

for $P\in P_n$, $|z|\geq 1$ and $R>1$, which contains Inequality (10) as a special case.

Remark 3. For $\alpha=0$, Inequality (12) yields,

$$\left|B[P(Rz)]+\beta\left(\frac{R+1}{r+1}\right)^n B[P(rz)]\right|$$
$$\leq\left|R^n+\beta\left(\frac{R+1}{r+1}\right)r^n\right|\left|B[z^n]\right|\max_{|z|=1}|P(z)| \tag{15}$$

for $|z|\geq 1,\ R>r\geq 1$ and $|\beta|\leq 1$.

If we choose $\lambda_0=\lambda_2=0$ in (12) and noting that all the zeros of $u(z)$ defined by (8) lie in the half plane (9), we get:

Corollary 1. If $P\in P_n$, then for all real or complex numbers α and β with $|\alpha|\leq 1$, $|\beta|\leq 1$, $R>r\geq 1$ and $|z|\geq 1$,

$$\left|RP'(Rz)+\phi(R,r,\alpha,\beta)rP'(rz)\right|$$
$$\leq n\left|R^n+\phi(R,r,\alpha,\beta)r^n\right|\left|z\right|^{n-1}\max_{|z|=1}\left|P(z)\right|. \quad (16)$$

where $\phi(R,r,\alpha,\beta)$ is defined as in Theorem 1. The result is sharp and equality in (16) holds for $P(z)=\lambda z^n$, $\lambda \neq 0.$

For the case $B[P(z)]=P(z)$, from (12) we obatin for all real or complex numbers α and β with $|\alpha|\leq 1$, $|\beta|\leq 1$, $R>r\geq 1$ and $|z|\geq 1$,

$$\left|P(Rz)+\phi(R,r,\alpha,\beta)P(rz)\right|$$
$$\leq\left|R^n+\phi(R,r,\alpha,\beta)r^n\right|\left|z\right|^{n}\max_{|z|=1}\left|P(z)\right|. \quad (17)$$

Inequality (17) is equivalent to the Inequality (5) for $|z|\geq 1$ and $\beta=0$. For $\alpha=0$ and $\beta=0$, Inequality (17) includes Inequality (2) as a special case.

Next we use Theorem 1 to prove the following result.

Theorem 2. If $P\in P_n$, then for arbitrary real or complex numbers α and β with $|\alpha|\leq 1$, $|\beta|\leq 1$, $R>r\geq 1$ and $|z|\geq 1$,

$$\begin{aligned}&\left|B\left[P(Rz)\right]+\phi(R,r,\alpha,\beta)B\left[P(rz)\right]\right|\\&\quad+\left|B\left[Q(Rz)\right]+\phi(R,r,\alpha,\beta)B\left[Q(rz)\right]\right|\\&\le\left\{\left|R^{n}+\phi(R,r,\alpha,\beta)r^{n}\right|\left|B\left[z^{n}\right]\right|\right.\\&\quad\left.+\left|1+\phi(R,r,\alpha,\beta)\right|\left|\lambda_0\right|\right\}\max_{|z|=1}\left|P(z)\right|\end{aligned}\tag{18}$$

where $Q(z)=z^{n}\overline{P(1/\overline{z})}$ and $\phi(R,r,\alpha,\beta)$ is defined as in Theorem 1.

The result is sharp and equality in (18) holds for $P(z)=\lambda z^{n}$, $\lambda\neq 0.$

Remark 4. Theorem 2 includes some well known polynomial inequalities as special cases. For example, inequality (18) reduces to a result due to Q. I. Rahman ([8], Inequality (5.2) with $\alpha=0$ and $\beta=0$). Also for $\beta=0$, Inequality (18) gives

$$\begin{aligned}&\left|B\left[P(Rz)\right]-\alpha B\left[P(rz)\right]\right|+\left|B\left[Q(Rz)\right]-\alpha B\left[Q(rz)\right]\right|\\&\le\left\{\left|R^{n}-\alpha r^{n}\right|\left|B\left[z^{n}\right]\right|+\left|1-\alpha\right|\left|\lambda_0\right|\right\}\max_{|z|=1}\left|P(z)\right|,\end{aligned}\tag{19}$$

for $|\alpha|\le 1$, $R>r\ge 1$ and $|z|\ge 1$.

If we take $\lambda_0=\lambda_2=0$ in (18), we get:

Corollary 2. If $P\in P_n$, then for all real or complex numbers α and β with $|\alpha|\le 1$, $|\beta|\le 1$, $R>r\ge 1$ and $|z|\ge 1$,

$$\begin{aligned}&\left|RP'(Rz)+\phi(R,r,\alpha,\beta)rP'(rz)\right|\\&\quad+\left|RQ'(Rz)+\phi(R,r,\alpha,\beta)rQ'(rz)\right|\\&\le n\left|R^n+\phi(R,r,\alpha,\beta)r^n\right|\left|z\right|^{n-1}\max_{|z|=1}\left|P(z)\right|.\end{aligned} \tag{20}$$

where $\phi(R,r,\alpha,\beta)$ is defined as in Theorem 1. The result is sharp and equality in (20) holds for $P(z)=\lambda z^n$, $\lambda\neq 0.$

For $\lambda_1=\lambda_2=0$ and $\alpha=1$, $\beta=0$, Theorem 2 includes a result due to A. Aziz and Rather [2] as a special case.

Inequality (12) can be sharpened if we restrict ourselves to the class of polynomials $P\in P_n$ having no zeros in $|z|<1$. In this direction we next prove the following result which is a compact generalization of the Inequalities (3), (4) and (6).

Theorem 3. If $P\in P_n$ and $P(z)\neq 0$ for $|z|<1$, then for arbitrary real or complex numbers α and β with $|\alpha|\le 1$, $|\beta|\le 1$, $R>r\ge 1$ and $|z|\ge 1$,

$$\begin{aligned}&\left|B\left[P(Rz)\right]+\phi(R,r,\alpha,\beta)B\left[P(rz)\right]\right|\\&\le\frac{1}{2}\left\{\left|R^n+\varphi(R,r,\alpha,\beta)r^n\right|\left|B\left[z^n\right]\right|\right.\\&\quad\left.+\left|1+\varphi(R,r,\alpha,\beta)\right|\left|\lambda_0\right|\right\}\max_{|z|=1}\left|P(z)\right|\end{aligned} \tag{21}$$

where $\phi(R,r,\alpha,\beta)$ is defined as in Theorem 1. The result is sharp and equality in (21) holds for $P(z)=z^n+1.$

Remark 5. Inequality (11) is a special case of the Inequality (21) for $\alpha=0$ and $\beta=0$. If we choose $\lambda_0=\lambda_2=0$ in (21) and note that all the zeros of $u(z)$ defined by (8) lie in the half plane defined by (9), it follows that if $P\in P_n$ and $P(z)\neq 0$ for $|z|<1$, then for $|z|\geq 1$, $R>r\geq 1$ and $|\alpha|\leq 1$, $|\beta|\leq 1$,

$$\left|RP'(Rz)+\phi(R,r,\alpha,\beta)rP'(rz)\right|$$
$$\leq\frac{n}{2}\left|R^n+\phi(R,r,\alpha,\beta)r^n\right||z|^{n-1}\max_{|z|=1}|P(z)|. \tag{22}$$

Setting $\alpha=0$ in (22), we obtain for $P\in P_n$,

$$\left|RP'(Rz)+\beta\left(\frac{R+1}{r+1}\right)^n rP'(rz)\right|$$
$$\leq\frac{n}{2}\left|R^n+\beta\left(\frac{R+1}{r+1}\right)^n r^n\right||z|^{n-1}\max_{|z|=1}|P(z)| \tag{23}$$

for $|z|\geq 1$, $R>r\geq 1$ and $|\beta|\leq 1$.

Taking $\alpha=\beta=0$ in (22), we obtain for $P\in P_n$, $|z|\geq 1$ and $R>1$,

$$\left|P'(Rz)\right| \le \frac{n}{2} R^{n-1} |z|^{n-1} \max_{|z|=1} \left|P(z)\right|, \tag{24}$$

which in particular gives Inequality (3).

Next choosing $\lambda_1 = \lambda_2 = 0$ in (21), we immediately get for $|z| \ge 1$, $R > r \ge 1$ and $|\alpha| \le 1$, $|\beta| \le 1$,

$$\begin{aligned} &\left|P(Rz) + \phi(R,r,\alpha,\beta)P(rz)\right| \\ &\le \frac{1}{2}\Big\{\left|R^n + \phi(R,r,\alpha,\beta)r^n\right||z|^n \\ &\quad + \left|1 + \phi(R,r,\alpha,\beta)\right|\Big\} \max_{|z|=1} \left|P(z)\right|, \end{aligned} \tag{25}$$

which is a compact generalization of the Inequalities (3), (4) and (6). The result is sharp and equality in (25) holds for $P(z) = az^n + b$, $|a| = |b| = 1$.

If we put $\beta = 0$ in (25), we get the following result.

Corollary 3. If $P \in P_n$, and $P(z) \ne 0$ for $|z| < 1$, then for every real or complex number α with $|\alpha| \le 1$, $R > r \ge 1$ and $|z| \ge 1$,

$$\begin{aligned} &\left|P(Rz) - \alpha P(rz)\right| \\ &\le \frac{1}{2}\Big\{\left|R^n - \alpha r^n\right||z|^n + \left|1 - \alpha\right|\Big\} \max_{|z|=1} \left|P(z)\right|. \end{aligned} \tag{26}$$

A polynomial $P \in P_n$ is said to be self-inversive if

$P(z)=Q(z)$ where $Q(z)=z^n\overline{P(1/\bar{z})}$. It is known [610] that if $P\in P_n$ is a self-inversive polynomial, then

$$\max_{|z|=1}|P'(z)|\le\frac{n}{2}\max_{|z|=1}|P(z)|. \tag{27}$$

Here finally, we establish the following result for self-inversive polynomials Theorem 4. If $P\in P_n$ is a self-inversive polynomial, then for arbitrary real or complex numbers α and β with $|\alpha|\le 1$, $|\beta|\le 1$, $R>r\ge 1$ and $|z|\ge 1$,

$$\begin{aligned}&|B[P(Rz)]+\phi(R,r,\alpha,\beta)B[P(rz)]|\\&\le\frac{1}{2}\{|R^n+\phi(R,r,\alpha,\beta)r^n||B[z^n]|\\&+|1+\phi(R,r,\alpha,\beta)||\lambda_0|\}\max_{|z|=1}|P(z)|\end{aligned} \tag{28}$$

where $\phi(R,r,\alpha,\beta)$ is defined as in Theorem 1. The result is sharp and equality in (21) holds for $P(z)=z^n+1$.

The following result is an immediate consequence of Theorem 4.

Corollary 4. If $P\in P_n$ is a self-inversive polynomial, then for arbitrary real or complex numbers α and β with $|\alpha|\le 1$, $|\beta|\le 1$, $R>r\ge 1$ and $|z|\ge 1$,

$$\begin{aligned}&\left|P(Rz)+\phi(R,r,\alpha,\beta)P(rz)\right|\\&\leq\frac{1}{2}\left\{\left|R^n+\phi(R,r,\alpha,\beta)r^n\right||z|^n\right.\\&\left.+\left|1+\phi(R,r,\alpha,\beta)\right|\right\}\max_{|z|=1}|P(z)|.\end{aligned} \tag{29}$$

where $\phi(R,r,\alpha,\beta)$ is defined as in Theorem 1. The result is best possible For $\beta=0$ the Inequality (29) reduces to

$$\begin{aligned}&\left|P(Rz)-\alpha P(rz)\right|\\&\leq\frac{1}{2}\left\{\left|R^n-\alpha r^n\right||z|^n+|1-\alpha|\right\}\max_{|z|=1}|P(z)|.\end{aligned} \tag{30}$$

Remark 6. Inequality (6) is a special case of the Inequality (30). Many other interesting results can be deduced from Theorem 4 in the same way as we have deduced from Theorem 1 and Theorem.

2. LEMMAS

For the proofs of these theorems, we need the following lemmas. The first lemma can be easily proved.

Lemma 1. If $P\in P_n$ and $P(z)$ has all its zeros in $|z|\leq 1$, then for every $R>r\geq 1$ and $|z|=1$,

$$|P(Rz)|\geq\left(\frac{R+1}{r+1}\right)^n|P(rz)|. \tag{31}$$

The next Lemma follows from corollary 18.3 of [11, p. 65].

Lemma 2. If $P \in P_n$ and $P(z)$ has all its zeros in $|z| \leq 1$, then all the zeros of $B[P(z)]$ also lie in $|z| \leq 1$.

Lemma 3. If $P \in P_n$ and $P(z)$ does not vanish in $|z| < 1$, then for arbitrary real or complex numbers α and β with $|\alpha| \leq 1$, $|\beta| \leq 1$, $R > r \geq 1$ and $|z| = 1$,

$$\begin{aligned}&\left|B[P(Rz)] + \phi(R,r,\alpha,\beta)B[P(rz)]\right| \\ &\leq \left|B[Q(Rz)] + \phi(R,r,\alpha,\beta)B[Q(rz)]\right|\end{aligned} \tag{32}$$

where $Q(z) = z^n \overline{P(1/\overline{z})}$ and $\phi(R,r,\alpha,\beta)$ is defined as in Theorem 1.

The result is sharp and equality in (32) holds for $P(z) = z^n + 1.$

Proof of Lemma 3. Since the polynomial $P(z)$ has all its zeros in $|z| \geq 1$ for every real or complex number γ with $|\gamma| > 1$, the polynomial $g(z) = P(z) - \gamma Q(z)$ where $Q(z) = z^n \overline{P(1/\overline{z})}$, has all its zeros in $|z| \geq 1$ with atleast one zero in $|z| < 1$, so that we can write

$$g(z) = \left(z - te^{i\delta}\right)h(z),$$

where $t<1$ and $h(z)$ is a polynomial of degree $n-1$ having all its zeros in $|z|\leq 1$.

Applying lemma 1 to the polynomial $h(z)$, we obtain for $R>r\geq 1$ and $0\leq\theta<2\pi$,

$$\left|g\left(\mathrm{Re}^{i\theta}\right)\right|=\left|\mathrm{Re}^{i\theta}-te^{i\delta}\right|\left|h\left(\mathrm{Re}^{i\theta}\right)\right|$$
$$\geq\left|\mathrm{Re}^{i\theta}-te^{i\delta}\right|\left(\frac{R+1}{r+1}\right)^{n-1}\left|h\left(re^{i\theta}\right)\right|$$

This implies for $R>r\geq 1$ and $0\leq\theta<2\pi$,

$$\left(\frac{r+t}{R+t}\right)\left|g\left(\mathrm{Re}^{i\theta}\right)\right|\geq\left(\frac{R+1}{r+1}\right)^{n-1}\left|g\left(re^{i\theta}\right)\right|. \tag{33}$$

Since $R>r\geq 1>t$ so that $g\left(\mathrm{Re}^{i\theta}\right)\neq 0$ for $0\leq\theta<2\pi$ and $\frac{1+r}{1+R}>\frac{r+t}{R+t}$, from Inequality (33), we obtain for $R>r\geq 1$ and $0\leq\theta<2\pi$,

$$\left|g\left(\mathrm{Re}^{i\theta}\right)\right|>\left(\frac{R+1}{r+1}\right)^{n}\left|g\left(re^{i\theta}\right)\right|. \tag{34}$$

Equivalently,

$$\left|g\left(Rz\right)\right|>\left(\frac{R+1}{r+1}\right)^{n}\left|g\left(rz\right)\right|$$

for $|z|=1$ and $R>r\geq 1$. Hence for every real or complex number α with $|\alpha|\leq 1$ and $R>r\geq 1$ we have

$$|g(Rz)-\alpha g(rz)|\geq |g(Rz)|-|\alpha||g(rz)|$$
$$>\left\{\left(\frac{R+1}{r+1}\right)^n-|\alpha|\right\}|g(rz)| \tag{35}$$

for $|z|=1$. Also, Inequality (34) can be written as

$$|g(re^{i\theta})|<\left(\frac{r+1}{R+1}\right)^n|g(\mathrm{Re}^{i\theta})| \tag{36}$$

for every $R>r\geq 1$ and $0\leq\theta<2\pi$ Since $g(\mathrm{Re}^{i\theta})\neq 0$ and $\left(\frac{r+1}{R+1}\right)^n<1$, from inequality (36), we obtain for $0\leq\theta<2\pi$ and $R>r\geq 1$,

$$|g(re^{i\theta})|<|g(\mathrm{Re}^{i\theta})|.$$

Equivalently,

$$|g(rz)|<|g(Rz)| \text{ for } |z|=1.$$

Since all the zeros of $g(Rz)$ lie in $|z|\leq (1/R)<1$, a direct application of Rouche's theorem shows that the polynomial $g(Rz)-\alpha g(rz)$ has all its zeros in $|z|<1$ for every real or complex number α with $|\alpha| \leq 1$. Applying Rouche's theorem again, it follows from (35) that for arbitrary real or complex numbers α, β with $|\alpha| \leq 1$, $|\beta| \leq 1$ and $R > r \geq 1$, all the zeros of the polynomial

$$\begin{aligned} F(z) &= g(Rz) + \varphi(R,r,\alpha,\beta) g(rz) \\ &= [\{P(Rz) - \gamma Q(Rz)\} \\ &\quad + \phi(R,r,\alpha,\beta)\{P(Rz) - \gamma Q(Rz)\}] \\ &= [P(Rz) + \phi(R,r,\alpha,\beta) P(rz)] \\ &\quad - \gamma [Q(Rz) + \phi(R,r,\alpha,\beta) Q(rz)] \end{aligned}$$

lie in $|z| < 1$ with $|\gamma| \geq 1$. Applying Lemma 2 to the polynomial $F(z)$ and noting that B is a linear operator, it follows that all the zeros of the polynomial

$$\begin{aligned} T(z) &= B[F(z)] \\ &= \{B[P(Rz)] + \phi(R,r,\alpha,\beta) B[P(rz)]\} \\ &\quad - \gamma \{B[Q(Rz)] + \phi(R,r,\alpha,\beta) B[Q(rz)]\} \end{aligned}$$

lie in $|z| < 1$ for all real or complex numbers α, β, γ with $|\alpha| \leq 1$, $|\beta| \leq 1$, $|\gamma| \leq 1$ and $R > r \geq 1$. This implies

$$\left|B[P(Rz)]+\phi(R,r,\alpha,\beta)B[P(rz)]\right|$$
$$\leq\left|B[Q(Rz)]+\phi(R,r,\alpha,\beta)B[Q(rz)]\right| \tag{37}$$

for $|z|\geq 1$, $|\alpha|\leq 1$, $|\beta|\leq 1$ and $R>r\geq 1$. If Inequality (38) is not true, then there is a point $z=w$ with $|w|\geq 1$ such that

$$\left|\{B[P(Rz)]+\phi(R,r,\alpha,\beta)B[P(rz)]\}_{z=w}\right|$$
$$>\left|\{B[Q(Rz)]+\phi(R,r,\alpha,\beta)B[Q(rz)]\}_{z=w}\right|.$$

But all the zeros of $Q(z)$ lie in $|z|\leq 1$, therefore, it follows (as in case of $g(z)$) that all the zeros of

$$Q(Rz)+\phi(R,r,\alpha,\beta)Q(rz)$$

lie in $|z|<1$. Hence by Lemma 2, all the zeros of

$$B[Q(Rz)]+\phi(R,r,\alpha,\beta)B[Q(rz)]$$

lie in $|z|<1$, so that

$$\{B[Q(Rz)]+\phi(R,r,\alpha,\beta)B[Q(rz)]\}_{z=w}\neq 0.$$

We take

$$\gamma = \frac{\left\{B\left[P(Rz)\right]+\phi(R,r,\alpha,\beta)B\left[P(rz)\right]\right\}_{z=w}}{\left\{B\left[Q(Rz)\right]+\phi(R,r,\alpha,\beta)B\left[Q(rz)\right]\right\}_{z=w}},$$

then γ is a well defined real or complex number with $|\gamma|>1$ and with this choice of γ, from (37) we obtain $T(w)=0$ where $|w|\geq 1$. This contradicts the fact that all the zeros of $T(z)$ lie in $|z|<1$. Thus

$$\left|B\left[P(Rz)\right]+\phi(R,r,\alpha,\beta)B\left[P(rz)\right]\right|$$
$$\leq\left|B\left[Q(Rz)\right]+\phi(R,r,\alpha,\beta)B\left[Q(rz)\right]\right|$$

for $|z|\geq 1$, $|\alpha|\leq 1$, $|\beta|\leq 1$ and $R>r\geq 1$. This proves (38) and hence Lemma 3.

3. PROOFS OF THE THEOREMS

Proof of Theorem 1. Let $M=\max_{|z|=1}|P(z)|$, then $|P(z)|\leq M$ for $|z|=1$. By Rouche's Theorem, it follows that all the zeros of the polynomial $H(z)=P(z)-\lambda Mz^n$ lie in $|z|<1$ for every real or complex number λ with $|\lambda|>1$, therefore, as before (as in Lemma 3), we conclude that all the zeros of the polynomial

$$G(z)=H(Rz)+\phi(R,r,\alpha,\beta)H(rz)$$

lie in $|z|<1$ for all real or complex numbers α and β with $|\alpha|\leq 1$ and $|\beta|\leq 1$. Hence by Lemma 2, the polynomial

$$\begin{aligned}T(z)&=B[G(z)]\\&=B[H(Rz)]+\phi(R,r,\alpha,\beta)B[H(rz)]\\&=B[P(Rz)]+\phi(R,r,\alpha,\beta)B[P(rz)]\\&\quad+\lambda[R^n+\phi(R,r,\alpha,\beta)r^n]B[z^n]M\end{aligned}$$

has all its zeros in $|z|<1$ for every real or complex number λ with $|\lambda|>1$. This implies for every real or complex numbers α and β with $|\alpha|\leq 1$, $|\beta|\leq 1$ and $R>r\geq 1$,

$$\begin{aligned}&|B[P(Rz)]+\phi(R,r,\alpha,\beta)B[P(rz)]|\\&\leq|R^n+\phi(R,r,\alpha,\beta)r^n||B[z^n]|M \text{ for } |z|\geq 1\end{aligned} \tag{38}$$

If Inequality (40) is not true, then there is a point $z=w$ with $|w|\geq 1$ such that

$$\begin{aligned}&|\{B[P(Rz)]+\phi(R,r,\alpha,\beta)B[P(rz)]\}_{z=w}|\\&>\{|R^n+\phi(R,r,\alpha,\beta)r^n||B[z^n]|_{z=w}M\}.\end{aligned}$$

Since $\{B[z^n]\}_{z=w}\neq 0$, we take

$$\lambda = \frac{\{B[P(Rz)] + \phi(R,r,\alpha,\beta)B[P(rz)]\}_{z=w}}{[R^n + \phi(R,r,\alpha,\beta)r^n]\{B[z^n]\}_{z=w} M},$$

so that λ is a well defined real or complex number with $|\lambda| > 1$ and with this choice of λ, from (39) we get $T(w) = 0$ where $|w| \geq 1$. This contradicts the fact that all the zeros of $T(z)$ lie in $|z| < 1$. Thus for every real or complex numbers α and β with $|\alpha| \leq 1$, $|\beta| \leq 1$ and $R > r \geq 1$,

$$\begin{aligned}&|B[P(Rz)] + \phi(R,r,\alpha,\beta)B[P(rz)]|\\ &\leq |R^n + \phi(R,r,\alpha,\beta)r^n||B[z^n]|M \text{ for } |z| \geq 1.\end{aligned}$$

This completes the proof of Theorem 1.

Proof of Theorem 2. Let $M = \max_{|z|=1}|P(z)|$, then $|P(z)| \leq M$ for $|z| = 1$. If μ is any real or complex number with $|\mu| > 1$, then by Rouche's Theorem, the polynomial $f(z) - \mu M$ does not vanish in $|z| < 1$. Applying Lemma 3 to the polynomial $f(z)$ and using the fact that B is a linear operator, it follows that for all real or complex numbers α and β with $|\alpha| \leq 1$, $|\beta| \leq 1$, $R > r \geq 1$ and for $|z| \geq 1$

$$\begin{aligned}&\left|B\left[f(Rz)\right]+\phi(R,r,\alpha,\beta)B\left[f(rz)\right]\right|\\&\leq\left|B\left[f^{*}(Rz)\right]+\phi(R,r,\alpha,\beta)B\left[f^{*}(rz)\right]\right|\end{aligned}$$

where

$$\begin{aligned}f^{*}(z)&=z^{n}\overline{f(1/\overline{z})}=z^{n}\overline{P(1/\overline{z})}-\overline{\mu}Mz^{n}\\&=Q(z)-\overline{\mu}Mz^{n},\end{aligned}$$

$Q(z)=z^{n}\overline{P(1/\overline{z})}$. Using the fact that $B[1]=\lambda_0$, we obtain

$$\begin{aligned}&\left|\left(B\left[P(Rz)\right]+\phi(R,r,\alpha,\beta)B\left[P(rz)\right]\right)\right.\\&\quad\left.-\mu\left(1+\phi(R,r,\alpha,\beta)\right)\lambda_0\right|\\&\leq\left|\left(B\left[Q(Rz)\right]+\phi(R,r,\alpha,\beta)B\left[Q(rz)\right]\right)\right.\\&\quad\left.-\overline{\mu}\left(R^{n}+\varphi(R,r,\alpha,\beta)r^{n}\right)B\left[z^{n}\right]M\right|\end{aligned}$$

for all real or complex numbers α and β with $|\alpha|\leq 1$, $|\beta|\leq 1$, $R>r\geq 1$ and $|z|\geq 1$. Now choosing the argument of μ such that

$$\begin{aligned}&\left|\left(B\left[Q(Rz)\right]+\phi(R,r,\alpha,\beta)B\left[Q(rz)\right]\right)\right.\\&\quad\left.-\overline{\mu}\left(R^{n}+\phi(R,r,\alpha,\beta)r^{n}\right)B\left[z^{n}\right]M\right|\\&=|\mu|\left|R^{n}+\phi(R,r,\alpha,\beta)r^{n}\right|\left|B\left[z^{n}\right]\right|M\\&\quad-\left|B\left[Q(Rz)\right]+\phi(R,r,\alpha,\beta)B\left[Q(rz)\right]\right|\end{aligned}$$

which is possible by Theorem 1, we get

$$\begin{aligned}&\left|B\left[P(Rz)\right]+\phi(R,r,\alpha,\beta)B\left[P(rz)\right]\right|\\&\quad-|\mu||1+\phi(R,r,\alpha,\beta)||\lambda_0|M\\&\le|\mu||R^n+\phi(R,r,\alpha,\beta)r^n||B[z^n]|M\\&\quad-\left|B\left[Q(Rz)\right]+\phi(R,r,\alpha,\beta)B\left[Q(rz)\right]\right|\end{aligned}$$

for $|\alpha|\le1$, $|\beta|\le1$, $R>r\ge1$ and $|z|\ge1$. This implies

$$\begin{aligned}&\left|B\left[P(Rz)\right]+\phi(R,r,\alpha,\beta)B\left[P(rz)\right]\right|\\&\quad+\left|B\left[Q(Rz)\right]+\phi(R,r,\alpha,\beta)B\left[Q(rz)\right]\right|\\&\le|\mu|\{|R^n+\phi(R,r,\alpha,\beta)r^n||B[z^n]|\\&\quad+|\lambda_0||1+\phi(R,r,\alpha,\beta)|\}M,\end{aligned}$$

for $|\alpha|\le1$, $|\beta|\le1$, $R>r\ge1$ and $|z|\ge1$. Letting $|\mu|\to1$, we obtain

$$\begin{aligned}&\left|B\left[P(Rz)\right]+\phi(R,r,\alpha,\beta)B\left[P(rz)\right]\right|\\&\quad+\left|B\left[Q(Rz)\right]+\phi(R,r,\alpha,\beta)B\left[Q(rz)\right]\right|\\&\le\{|R^n+\phi(R,r,\alpha,\beta)r^n||B[z^n]|\\&\quad+|\lambda_0||1+\phi(R,r,\alpha,\beta)|\}M,\end{aligned}$$

which is inequality (18) and this proves Theorem 2.

Proof of Theorem 3. Lemma 3 and Theorem 2 together yields for all real or complex numbers α and β with $|\alpha|\le 1$, $|\beta|\le 1$, $R>r\ge 1$ and $|z|\ge 1$,

$$\begin{aligned}
&2\left|B\left[P(Rz)\right]+\phi(R,r,\alpha,\beta)B\left[P(rz)\right]\right| \\
&=\left|B\left[P(Rz)\right]+\phi(R,r,\alpha,\beta)B\left[P(rz)\right]\right| \\
&\quad+\left|B\left[P(Rz)\right]+\phi(R,r,\alpha,\beta)B\left[P(rz)\right]\right| \\
&\le\left|B\left[P(Rz)\right]+\phi(R,r,\alpha,\beta)B\left[P(rz)\right]\right| \\
&\quad+\left|B\left[Q(Rz)\right]+\phi(R,r,\alpha,\beta)B\left[Q(rz)\right]\right| \\
&\le\left\{\left|R^n+\phi(R,r,\alpha,\beta)r^n\right|\left|B\left[z^n\right]\right|\right. \\
&\quad\left.+\left|1+\phi(R,r,\alpha,\beta)\right|\right\}M,
\end{aligned}$$

which gives

$$\begin{aligned}
&\left|B\left[P(Rz)\right]+\phi(R,r,\alpha,\beta)B\left[P(rz)\right]\right| \\
&\le\frac{1}{2}\left\{\left|R^n+\phi(R,r,\alpha,\beta)r^n\right|\left|B\left[z^n\right]\right|\right. \\
&\quad\left.+\left|\lambda_0\right|\left|1+\phi(R,r,\alpha,\beta)\right|\right\}M,
\end{aligned}$$

which is the Inequality (21) and this completes the proof of Theorem 3.

Proof of Theorem 4. Since $P(z)$ is a self-inversive polynomial of degree n, therefore

$$P(z)=Q(z)=z^n\overline{P(1/\overline{z})}$$

for all $z \in C$. This implies, in particular, that for all real or complex numbers α and β with $|\alpha| \le 1$, $|\beta| \le 1$, $R > r \ge 1$ and $|z| \ge 1$,

$$\begin{aligned} &\left|B[P(Rz)] + \phi(R,r,\alpha,\beta)B[P(rz)]\right| \\ &= \left|B[Q(Rz)] + \phi(R,r,\alpha,\beta)B[Q(rz)]\right|. \end{aligned}$$

Combining this with Theorem 2, the desired result follows immediately. This completes the proof of Theorem 4.

4. ACKNOWLEDGEMENTS

Authors are thankful to the referee for his suggestions.

REFERENCES

1. G. V. Milovanovic, D. S. Mitrinovic and Th. M. Rassias, "Topics in Polynomials: Extremal Properties, Inequalities, Zeros," World scientific Publishing Co., Singapore, 1994.

2. Q. I. Rahman and G. Schmessier, "Analytic Theory of Polynomials," Claredon Press, Oxford,2002.

3. A. C. Schaffer, "Inequalities of A. Markoff and S. Bernstein for Polynomials and Related Functions," Bulletin of the American Mathematical Society, Vol. 47, No. 2, 1941, pp. 565-579.

4. M. Riesz, "Uber Einen Satz des Herrn Serge Bernstein," Acta Mathematica, Vol. 40, 1916, pp. 337-347.

5. G. Pólya and G. Szegö, "Aufgaben und Lehrsätze aus der Analysis," Springer-Verlag, Berlin, 1925.

6. P. D. Lax, "Proof of a Conjecture of P. Erdös on the Derivative of a Polynomial," Bulletin of the American Mathematical Society, Vol. 50, No. 8, 1944, pp. 509-513.

7. N. C. Ankeny and T. J. Rivlin, "On a Theorm of S. Bernstein," Pacific Journal of Mathematics, Vol. 5, 1955, pp. 849-852.

8. A. Aziz and N. A. Rather, "On an Inequality of S. Bernstein and Gauss-Lucas Theorem," Analytic and Geometriv Inequalities, Kluwer Academic Pub., Dordrecht, 1999, 29-35.

9. Q. I. Rahman, "Functions of Exponential Type," Transactions of the American Mathematical Society, Vol. 135, 1969, pp. 295-309.

10. E. B. Saff and T. Sheil-Small, "Coefficient and Integral Mean Estimates for Algebraic and Trigonometric Polynomials with Restricted Zeros," Journal of the London Mathematical Society, Vol. 9, No. 2, 1975, pp. 16-22

11. M. Marden, "Geometry of Polynomials," Surveys in Mathematics, No. 3, 1949.

14

CHAPTER

Programming First Integral Method General Formula for the Solving Linear and Nonlinear Equations

Mohamed A. Abdoon

Mathematics Department, Sudan University of Science and Technology, Khortom, Sudan

ABSTRACT

It's well known that the solution of equations always uses complicated methods. In this paper the first integral method is used to find the actual solution of equations in a simple way, rather than the ex-complicated ways. Therefore, the use of first integral method makes the solution more available and easy to investigate behavior waves through its solution. First integral method is used to find exact solutions to the general formula and the applications of the results to the linear and nonlinear equations.

Keywords: First Integral Method, Exact Solution, Linear and Nonlinear Equations

1. INTRODUCTION

Partial differential equations arise frequently in the formulation of fundamental laws of nature and in the mathematical analysis of a wide variety of problems in applied mathematics, mathematical physics, and engineering science. This subject plays a central role in modern mathematical sciences, especially in physics, geometry, and analysis. Many problems of physical interest are described by partial differential equations with appropriate initial and/or boundary conditions. These problems are usually formulated as initial-value problems, boundary- value problems, or initial boundary-value problems, a broad coverage of the essential standard material on linear partial differential equations and their applications is required. The study of the solutions of partial differential equations (PDEs) has enjoyed an intense period of activity over the last forty years from both theoretical and numerical points of view. Many methods obtaining the exact solution of non linear equation, some of the techniques are the bilinear transformation [1] , the sine cosine method [2] , F-expansion method [3] , the first integral method was first proposed by Feng [4] to solving Burger-Korteweg-devries equation and so on, in this paper investigation a traveling wave solution for non linear partial differential equation, study nonlinear phenomena, in solving modified KdV-kp can be based on the theory of commutative algebra, using the first integral method technique to solving linear and nonlinear equations.

2. FIRST INTEGRAL METHOD

The non-linear partial differential equation form:

$$w\left(F, F_x, F_t, F_{xx}, F_{xt}, \ldots\right) \tag{1}$$

where $u(x,t)$ is the solution of (1) we use the transforms:

$$f(x,t) = f(\zeta), \zeta = \alpha x - \beta t \tag{2}$$

we use the wave transforms :

$$\frac{\partial}{\partial t}(.)=-\beta\frac{\partial}{\partial\zeta}(.),\ \frac{\partial}{\partial x}(.)=\alpha\frac{\partial}{\partial\zeta}(.),\ \frac{\partial^2}{\partial t^2}(.)=\beta^2\frac{\partial^2}{\partial\zeta^2}(.),\ \frac{\partial^2}{\partial x^2}(.)=\alpha^2\frac{\partial^2}{\partial\zeta^2}(.) \tag{3}$$

Equation (1) transforms the ordinary differential equations we obtain:

$$p(f,f',f'',\text{L})=0 \tag{4}$$

Anew independent variable:

$$x(\zeta)=f(\zeta),\ y(\zeta)=f_\zeta(\zeta) \tag{5}$$

The system of ordinary differential equations:

$$\begin{aligned} x'(\zeta)&=y(\zeta)\\ y'(\zeta)&=F(x(\zeta),y(\zeta)) \end{aligned} \tag{6}$$

By the qualitative theory of differential equation [5] , we find the integral of (6) under same condition, then the general solution of (6) can be obtained directly. However, in general, it is really difficult for us to realize this even for one first integral, because for a given plane autonomous system, find its first integral will apply the Division theory to option first integral (6), An exact solution of (1) obtained by solving this equation. Now let us recall the Division theory.

Division theorem:

Suppose that $P(x,y)$ and $Q(x,y)$ are polynomials of two variables x and y in $C[x,y]$. And $P(x,y)$ is irreducible in $C[x,y]$. If $Q(x,y)$ vanishes at all points of $P(x,y)$, then there exists a polynomial $G(X,Y)$ in $C[x,y]$ such that $Q(X,Y)=P(X,Y)G(X,Y)$.

3. THE FIRST INTEGRAL METHOD GENERAL FORMULA

We discuss the problem by using the first integral method, consider the general formula:

$$u''(\xi)-su'-k-lu(\xi)-mu^2(\xi)-nu^3(\xi)-fu^4(\xi)-ru^5(\xi)=0 \quad (7)$$

where s,k,l,m,n,f,r are real constant. Using (7) in (6) we get the system.

$$X_\xi(\xi)=Y(\xi),$$
$$Y_\xi(\xi)=sY+k+lX(\xi)+mX^2(\xi)+nX^3(\xi)+fX^4(\xi)+rX^5(\xi) \quad (8)$$

Now Appling Division theorem, suppose that $X(\zeta)$ and $Y(\zeta)$ are nontrivial solution of (8):

$$q(X,Y)=\sum_{i=0}^{M}a_i(X)Y^i=0 \quad (9)$$

Is an irreducible polynomial in the complex domain $C[X,Y]$ such that:

$$q[X(\zeta),Y(\zeta)]=\sum_{i=0}^{M}a_i(X(\zeta))Y^i(\zeta)=0 \quad (10)$$

$a_i(X)(i=0,1,2,\ldots,m)$ are polynomial and $a_m(X)\neq 0$, Equation (16) called first integral method, there exist a polynomial $g(X)X+h(X)Y$ in the complex domain $C[x,y]$ such that:

$$\frac{dq}{d\zeta}=\frac{dq}{dX}\cdot\frac{dX}{d\zeta}+\frac{dq}{dY}\cdot\frac{dY}{d\zeta}=(g(X)X+h(X)Y)\sum_{i=0}^{M}a_i(X)Y^i \quad (11)$$

which can be written as:

$$\sum_{i=0}^{M} a_i^{\bullet}(X)Y^{i+1}+\sum_{i=0}^{M} ia_i(X)Y^{i-1}\left(k+lX+mX^2+nX^3+fX^4+rX^5\right)=\left(g(X)+h(X)Y\right)\sum_{i=0}^{M} a_i(X)Y^i \quad (12)$$

by comparing with the coefficient of $Y^i\ (i=M+1,M,\text{L}\ ,1,0)$ on both sides of (12), we get:

$$a_M^{\bullet}=h(X)a_M(X) \quad (13a)$$

$$a_{M-1}^{\bullet}=g(X)a_M(X)+h(X)a_{M-1}(X) \quad (13b)$$

$$a_0^{\bullet}(X)=-2a_2(X)\left(k+lX+mX^2+nX^3+fX^4+rX^5\right)=g(X)a_0(X) \quad (13x)$$

$$a_1(X)\left(k+lX+mX^2+nX^3+fX^4+rX^5\right)=g(X)a_0(X) \quad (13y)$$

from (12a), we deduce that $a_m(X)$ is a constant and $h(X)=0$, we take $a_m(X)=1$, and balancing the degrees of $g(X), a_1(X)$ and $a_0(X)$, we find the deg g(X).

Now we take these cases:

Case 1:

Suppose that M = 1, in (12), then the (13) becomes:

$$a_1^{\bullet}(X)=h(X)a_1(X) \quad (14a)$$

$$a_0^{\bullet}(X)+sa_1(X)=g(X)a_1(X)+h(X)a_0(X) \quad (14b)$$

$$a_1(X)\left(k+lX+mX^2+nX^3+fX^4+rX^5\right)=g(X)a_0(X) \quad (14c)$$

since $a_i(X)(i=0,1)$ are polynomial, then from (14a) we deduce that $a_1(X)$ is constant and $h(X)=0$ for simplicity, take $a_1(X)=1$. Balancing the degrees of $g(X)$ and $a_0(X)$. We conclude that $\deg(g(X))=2$, suppose that $g(X)=A_2X^2+A_1X+A_0$, then we find $a_0(X)$.

$$a_0(X)=B_0+(A_0-s)X+\frac{A_1X^2}{2}+\frac{A_2X^3}{3} \tag{15}$$

where B_0 is arbitrary integration constant. Substituting $a_0(X), a_1(X)$ and $g(X)$ in (14c), and setting all the coefficients of powers X to be zero, we obtain a system of nonlinear algebraic equations and by solving it, we obtain:

$$r=\frac{1}{3}A_2^2, f=\frac{5}{6}A_1A_2, n=\frac{4}{3}A_2A_0+\frac{A_1^2}{2}-sA_2, m=\frac{3}{2}A_0A_1-sA_1, l=A_0^2-sA_0, k=0, B_0=0 \tag{16}$$

using (16) in (10), we obtain:

$$Y=\frac{-2A_2X^3-3A_1X^2-6(A_0-s)X}{6} \tag{17}$$

combining (17) with (8), and fine the exact solution (8).

Case 2:

Suppose that M = 2, in (12), then the (13) became:

$$a_2^{\bullet}(X)=h(X)a_2(X) \tag{18a}$$

$$a_1^{\bullet}(X)+2a_2(X)s=g(X)a_2(X)+h(X)a_1(X) \tag{18b}$$

$$a_0^{\bullet}(X)+a_1(X)s+2a_2\left(k+lX+mX^2+nX^3+fX^4+rX^5\right)=g(X)a_1(X)+h(X)a_0(X) \quad (18c)$$

$$a_1(X)\left(k+lX+mX^2+nX^3+fX^4+rX^5\right)=g(X)a_0(X) \quad (18d)$$

since $a_2(X)(i=0,1,2)$ are polynomial, then from (14a) we deduce that $a_2(X)$ is constant and $h(X)=0$ for simplicity, take $a_2(X)=1$. Balancing the degrees of $g(X), a_1(X)$ and $a_0(X)$. We conclude that $\deg(g(X))=2$, suppose that $g(X)=A_2X^2+A_1X+A_0$, then we find $a_1(X)$ and $a_0(X)$:

$$a_1(X)=B_0+(A_0-2s)X+\frac{A_1X^2}{2}+\frac{A_2X^3}{3} \quad (19)$$

$$a_0(X)=B_1+[A_0B_0-2k]X+\frac{1}{2}\left[A_1B_0+A_0^2-3A_0s+2s^2-2l\right]X^2+\frac{1}{3}\left[A_2B_0+\frac{3}{2}A_0A_1-\frac{5A_1s}{2}-2m\right]X^3$$
$$+\frac{1}{4}\left[\frac{4}{3}A_0A_2+2A_2s+\frac{A_1^2}{2}-2n\right]X^4+\frac{1}{5}\left[\frac{5}{6}A_1A_2-2f\right]X^5+\frac{1}{6}\left[\frac{A_2^2}{3}-2r\right]X^6 \quad (20)$$

where B_0, B_1 are arbitrary integration constants. Substituting $a_0(X)\cdot a_1(X)$ and $g(X)$ in (18d), and setting all the coefficients of powers X to be zero, we obtain a system of nonlinear algebraic equations and by solving it, we obtain:

$$k=0, l=\frac{A_0^2}{4}-\frac{sA_0}{2}, m=\frac{9A_0A_1}{16}-sA_1, n=\frac{4A_0A_2+A_1^2}{8}+\frac{7sA_2}{5}, f=\frac{5A_1A_2}{16}, r=\frac{A_2^2}{12}, B_1=0, B_0=0 \quad (21)$$

using (20) in (9),we obtain two equal roots for Y:

note that:

$$a_0(X)=\frac{a_1^2(X)}{4},\ \text{then } Y=\frac{-2A_2X^3-3A_1X^2-6(A_0-s)X}{12}, \quad (22)$$

combining (21) with (8), and fine the exact solution (8).

Case 3:

Suppose that M = 3, in (12), then the (13) became:

$$a_3^\bullet(X)=h(X)a_3(X) \quad (23a)$$

$$a_2^\bullet(X)+3sa_3(X)=g(X)a_3(X)+h(X)a_2(X) \quad (23b)$$

$$a_1^\bullet(X)+2sa_2(X)+3a_3(k+lX+mX^2+nX^3+fX^4+rX^5)=g(X)a_2(X)+h(X)a_1(X) \quad (23c)$$

$$a_0^\bullet(X)+sa_1(X)+2a_2(k+lX+mX^2+nX^3+fX^4+rX^5)=g(X)a_1(X)+h(X)a_0(X) \quad (23d)$$

$$a_1(X)(k+lX+mX^2+nX^3+fX^4+rX^5)=g(X)a_0(X) \quad (23e)$$

since $a_i(X)(i=0,1,2)$ are polynomial, then from (14a) we deduce that $a_2(X)$ is constant and $h(X)=0$ for simplicity, take $a_2(X)=1$. Balancing the degrees of $g(X),a_1(X)$ and $a_0(X)$. We conclude that $\deg(g(X))=2$, suppose that $g(X)=A_2X^2+A_1X+A_0$, then we find $a_2(X),a_1(X)$ and $a_0(X)$:

$$a_2(X) = B_0 + (A_0 - 3s)X + \frac{A_1 X^2}{2} + \frac{A_2 X^3}{3} \tag{24}$$

$$\begin{aligned} a_1(X) = {} & \frac{1}{6}\left[-3r + \frac{1}{3}A_2^2\right]X^6 + \frac{1}{5}\left[\frac{5}{6}A_1A_2 - 3f\right]X^5 + \frac{1}{4}\left[\frac{4}{3}A_0A_2 + \frac{1}{2}A_1^2 + \frac{sA_2}{3} - 3n\right]X^4 \\ & + \frac{1}{3}\left[\frac{3}{2}A_0A_1 - (s+3)A_1 - 3m\right]X^3 + \frac{1}{2}\left[A_0^2 + 6s^2 - 5sA_0 - 3l\right]X^2 \end{aligned} \tag{25}$$

$$\begin{aligned} a_0(X) = {} & B_1 + \frac{1}{9}\left[\frac{A_2^3}{18} - \frac{1}{6}A_2 r\right]X^9 + \frac{1}{8}\left[\frac{2}{9}A_1A_2^2 - \frac{4}{15}fA_2\right]X^8 \\ & + \frac{1}{7}\left[\frac{7}{18}A_0A_2^2 + \frac{7}{24}A_1^2A_2 - \frac{5}{12}nA_2 - \frac{1}{10}fA_1 + \frac{1}{2}rA_0\right]X^7 \\ & + \frac{1}{6}\left[\frac{1}{2}A_1A_2^2 - \frac{2}{3}mA_2 + \frac{1}{2}A_0A_1A_2 + \frac{1}{8}A_1^3 - \frac{1}{4}nA_1 + \frac{4}{5}A_0 f\right]X^6 \\ & + \frac{1}{5}\Bigg[\frac{5}{6}A_0^2A_2 + \frac{1}{2}A_2^2A_2 + \frac{1}{8}A_0A_1^2 - \frac{7}{6}lA_2 - \frac{1}{2}mA_1 \\ & + \frac{1}{4}nA_0 + \frac{25sA_0A_2}{6} + \frac{41s^2A_2}{12} + (9n - 4A_1)s\Bigg]X^5 \\ & + \frac{1}{4}\left[-2lA_1 - \frac{13A_0A_1}{3} + 2ms + \frac{4sA_1}{3} - \frac{lA_2}{3} + \frac{A_0^2A_1}{2}\right]X^4 \\ & + \frac{1}{3}\left[\frac{-5}{2}lA_0 + \frac{9ls}{2} - 3sA_0^2 + \frac{11s^2A_0}{2} + \frac{A_0^2}{2} - 3s^2\right]X^3 \end{aligned} \tag{26}$$

where B_0, B_1 are arbitrary integration constants.

Substituting $a_0(X)\cdot a_1(X)$ and $g(X)$ in (18d), and setting all the coefficients of powers X to be zero, we obtain a system of nonlinear algebraic equations and by solving it, we obtain:

$$k = 0, l = \frac{A_0^2}{4} + 3A_0 s - 4s^2, m = \frac{9A_0 A_1}{16} + \frac{8sA_1}{3},$$
$$n = \frac{4A_0 A_2 + A_1^2}{12} + \frac{A_2 s}{3}, f = \frac{5A_1 A_2}{16}, r = \frac{A_2^2}{12}, B_1 = 0, B_0 = 0 \quad (27)$$

using (20) in (9), we obtain three equal roots for Y:

note that:

$$a_1(X) = \frac{a_2^2(X)}{3}, a_0(X) = \frac{a_2^3(X)}{27},$$

then

$$Y = \frac{-2A_2 X^3 - 3A_1 X^2 - 6(A_0 - s)X}{18}, \quad (28)$$

combining (21) with (8), and fine the exact solution (8).

Case n:

Suppose M = n, we get:

$$Y = \frac{-2A_2 X^3 - 3A_1 X^2 - 6(A_0 - s)X}{6n} \quad (29)$$

Theorem 3.1:

The exact solution of the general formula in (7) are given by combining of (17), (21), (27) ... (29), with (8) and integration respect with ξ.

4. APPLICATION

We can apply Theorem 3.1 to studying some nonlinear differential equations, as solitary wave equation.

Example 4.1:

The linear ODES:

$$u'' - u = 0 \tag{30}$$

which is the same form of Equation (7), where:

$$l = 1, m = n = f = r = 0, s = 0$$

$$\begin{aligned} &A_0 = 1, A_1 = 0, A_2 = 0, \\ &A_0 = -1, A_1 = 0, A_2 = 0, \end{aligned} \tag{31}$$

$$Y = -A_0 x \tag{32}$$

integration respect to ξ :

$$u_1 = Ae^x \tag{33}$$

$$u_2 = Be^{-x} \tag{34}$$

so the (31) + (32) is the solution:

$$u = Ae^x + Be^{-x}$$

Example 4.2:

Consider the Boussines equation given by:

$$u_{tt}=u_{xx}+3\left(u^2\right)_{xx}+u_{xxxx}, a\le x\le b \tag{35}$$

using $u=f\left(x-ct\right)$ into (34) gives:

$$\left(c^2-1\right)f''=3\left(f^2\right)''+f^4 \tag{36}$$

where integrating twice yields:

$$f''=\left(c^2-1\right)f-3f^2 \tag{37}$$

which is the same form of (7), where:

$$l=\left(c^2-1\right), m=-3, n=f=r=0, s=0$$

$$A_0^2=\sqrt{c^2-1}, A_1=\frac{-2}{\sqrt{c^2-1}}, A_2=0 \tag{38}$$

$$Y=\frac{X^2-\left(c^2-1\right)X}{\sqrt{c^2-1}} \tag{39}$$

integration respect to ξ then:

$$u\left(x,t\right)=\frac{c^2-1}{2}\left[1+2\tanh\left(\sqrt{c^2-1}\right)\left[x-ct\right]\right] \tag{40}$$

Example 4.3:

Consider the Gardner equation given by:

$$u_t+2auu_x-3bu^2u_x+u_{xxx}=0, \left(a,b\right)\text{f } 0 \tag{41}$$

using the wave variable $\xi = x - ct$ and integrating the result will convert the (41) to the ODE:

$$f'' = bf^3 - af^2 + cf \tag{42}$$

which is the same form of (7) where:

$$l = c, m = -a, n = b, f = r = 0, s = 0$$

$A_0 = \sqrt{c}, A_1 = \sqrt{2b}, A_2 = 0$ and (43)

so: $Y = \dfrac{-\sqrt{2b}X^2 - 2\sqrt{c}X}{3}$ (44)

integrating respect to ξ then:

$$u(x,t) = \frac{1}{2}\sqrt{\frac{2c}{b}}\left[\tanh\left(\frac{-3}{2}\sqrt{\frac{2c}{b}}[x - ct] - 1\right)\right] \tag{45}$$

Example 4.4:

Consider the nonlinear Schrödinger equation:

$$iu_t + u_{xx} + \beta|u|^2 u + i\alpha\left(|u|^2 u\right)_x + is\left(|u|^2\right)_x u = 0 \tag{46}$$

suppose that (46) has solution form:

$$u(x,t) = e^{i(\phi(\eta) - wt)}\alpha(\eta), \eta = x - ct \tag{47}$$

substituting (47) in (46), then (46) become:

$$\alpha''(\eta)+\left(\frac{c^2}{4}+w\right)\alpha(\eta)+\left(\beta-\frac{\alpha c}{2}\right)\alpha^3(\eta)+\left(r+\frac{1}{16}(3\alpha+2s)(\alpha-2s)\right)\alpha^5(\eta)=0 \tag{48}$$

which is the same form of (48), where:

$$l=-\left(\frac{c^2}{4}+w\right),n=-\left(\beta-\frac{\alpha c}{2}\right),r=-\left(r+\frac{1}{16}(3\alpha+2s)(\alpha-2s)\right),m=0,f=0$$

$$A_2=\sqrt{3r},A_1=0,A_0=\frac{3n}{4\sqrt{3r}} \tag{49}$$

$$Y=\frac{-3A_0X-A_2X^3}{9} \tag{50}$$

integrating respect to ξ then:

$$u(x,t)=\frac{\pm\sqrt{3A_0}}{\sqrt{-A_2+e^{\frac{2}{3}A_0(\varsigma+\varsigma_0)}}} \tag{51}$$

Example 4.5:

The Cahn-Allen equation: we study nonlinear parabolic PDF given by:

$$u_t=u_{xx}-u^3+u \tag{52}$$

using the wave variable $\xi=x-ct$ and integrating the result will convert the (52) to the ODE:

$$-cu'=u''-u^3+u \tag{53}$$

which is the same form of (48), where:

$$k = f = m = 0, n = 1, l = -1, s = -c \tag{54}$$

$$A_0 = -\frac{1}{c}, A_1 = \pm\sqrt{2}, c = \pm\sqrt{3} \tag{55}$$

$$Y = \mp\frac{1}{\sqrt{2}}X^2 \pm \frac{2}{\sqrt{3}}X \tag{56}$$

integrating respect to ξ then:

$$u_1(x,t) = \frac{\frac{2}{\sqrt{3}}}{e^{-(\varsigma+\varsigma_0)} - \frac{\sqrt{3}}{2\sqrt{2}}} \tag{57}$$

$$u_2(x,t) = \frac{\frac{2}{\sqrt{3}}}{e^{-(\varsigma+\varsigma_0)} + \frac{\sqrt{3}}{2\sqrt{2}}} \tag{58}$$

$$u_3(x,t) = \frac{\frac{2}{\sqrt{3}}}{e^{(\varsigma+\varsigma_0)} - \frac{\sqrt{3}}{2\sqrt{2}}} \tag{59}$$

$$u_4(x,t) = \frac{\frac{2}{\sqrt{3}}}{e^{(\varsigma+\varsigma_0)} + \frac{\sqrt{3}}{2\sqrt{2}}} \tag{60}$$

our result can be compared to Wawaz's result [2] .

5. CONCLUSION

Mohamed A. Abdoon The first integral method a general formula, is successful for solving a lot of nonlinear equation, and establishing travelling wave solutions, which is based on the ring theory of commutative algebra, and used to solve complicated and tedious algebra calculation. We can also apply them to some other nonlinear partial differential equations.

REFERENCES

1. Hirota, R. (1980) Direct Method of Finding Exact Solutions of Nonlinear Evolution Equation. In: Bullongh, R. and Caudry, P., Eds., Backlund Transformation, Springer, Berlin, 115.

2. Wazwaz, A.M. (2004) A Sine-Cosine Method for Handling Nonlinear Wave Equations. Mathematical and Computer Modelling, 40, 499-508.

3. Zhou, Y.B., Wang, M.L. and Wang, Y.M. (2003) Periodic Wave Solutions to a Coupled KdV Equations with Variable Coefficients. Physics Letters A, 308, 31-36.

4. Feng, Z.S. (2002) The First Integral Method to Study the Burgers-Korteweg-de Vriesequation. Journal of Physics A, 35, 343-349.

5. Feng, Z. (2002) On Explicit Exact Solutions for the Lienard Equation and Its Applications. Physics Letters A, 293, 50-56.

INDEX